Journal of
Neural
Transmission

Supplementum 35

C. Marescaux, M. Vergnes, and R. Bernasconi (eds.)

Generalized
Non-Convulsive Epilepsy:
Focus on GABA-B Receptors

Springer-Verlag *Wien New York*

Prof. Dr. C. Marescaux
Clinique Neurologique, Hôpital Civil, Strasbourg, France

Dr. M. Vergnes
Centre de Neurochimie, C.N.R.S., Strasbourg, France

Dr. R. Bernasconi
Pharma Division, Ciba-Geigy AG, Basel, Switzerland

With 61 Figures

ISSN 0303-6995
ISBN-13:978-3-211-82340-8   e-ISBN-13:978-3-7091-9206-1
DOI: 10.1007/978-3-7091-9206-1

Preface

Absence or petit mal seizures and their electroencephalographic (EEG) correlate, the generalized, bilaterally synchronous spike-and-wave discharge (SWD), differ in many important respects from other forms of epileptic seizure. Behaviourally, the absence seizures are characterized by a brief interruption of responsiveness and activity, which are resumed at the end of the EEG discharge. They have been related to a predominance of inhibitory activity, in contrast to generalized or focal convulsive seizures where an excess of excitation is prevalent.

The EEG high-voltage SWD emerge from a normal background activity. The pattern appears to have a sudden simultaneous onset over regions of the scalp; it stops just as abruptly and simultaneously. In contrast to generalized tonic-clonic and partial seizures, it leaves no postictal depression after its termination. The generalized SWD shows no significant evolution in the course of an absence seizure, whereas in convulsive seizures the electrographic pattern evolves continuously. Given these characteristics of SWD, studies have been preferentially directed towards understanding abnormal interactions in widely distributed neuronal systems.

The pharmacological reactivity of absence seizures is also specific and distinct from that of convulsive epilepsy. Ethosuximide, a selective anti-absence drug, is ineffective against focal or convulsive seizures. Phenytoin and carbamazepine, which are classical anticonvulsants, aggravate the SWD in absence epilepsy.

The most extensively studied model of absence seizures is the penicillin-induced SWD in cats described by Gloor's group in Montreal. The purpose of the present studies was to approach the pathophysiology of absence seizures by way of various models of generalized non-convulsive epilepsy in rodents, and in particular genetic models which may be close to the typical absence epilepsy of childhood.

After a clinical introduction dealing with absence epilepsy in humans by Loiseau, different experimental models are described and discussed, viz: gamma-hydroxybutyrate-induced SWD in rats by Snead and genetic and spontaneous SWD in the mouse by Kostopoulos and in the rat by Marescaux et al. The following papers cover a range of topics concerned with the pathophysiology of absence epilepsy: cellular neurophysiology by Kostopoulos and Psarropoulou; analysis of the thalamocortical loop by Avanzini et al. and by Vergnes and Marescaux; cortical response to NMDA-stimulation by Pumain et al.; inhibitory control of the substantia nigra by Depaulis; metabolic investigations by Nehlig et al. Pharmacological studies reported by Bernasconi et al. and by Marescaux et al. revolve around GABA, and more specifically $GABA_B$ neurotransmission. Given the importance of $GABA_B$ activation or blockade for SWD, the GABA receptors were compared in epileptic and non-epileptic rats by Knight and Bowery.

Some of these results were reported in a session on generalized non-convulsive epilepsy during the 1990 European Winter Conference on Brain Research at Les Arcs, France, organized by S. Nicolaidis. This multidisciplinary approach to the understanding of a specific epileptic disorder should provide stimulus for the further research necessary to unravel the pathophysiology of petit mal or absence epilepsy.

We gratefully acknowledge financial support from Ciba-Geigy Ltd. We also appreciate the readiness of Springer-Verlag Wien New York to publish this special issue on the pathophysiological and biochemical mechanisms involved in absence epilepsy and the assistance of Alan Holmes Kirkwood in editorial matters.

CH. MARESCAUX
M. VERGNES
R. BERNASCONI

Contents

Listed in Current Contents

J Neural Transm (1992) [Suppl] 35: 1–6

Human absence epilepsies

P. Loiseau

Service de Neurologie, C.H.U., Hôpital Pellegrin, Bordeaux, France

Summary. A historical review of the concept of absence seizures. Their clinical features are very suggestive but a diagnosis made solely on clinical grounds is not always safe. Comparable pitfalls exist in the interpretation of EEG patterns. Absence seizures belong to several epileptic syndromes. They are briefly described.

The essential feature of absence epilepsies is absence seizures (AS). An AS is characterized by a sudden interruption of activity, a vacant stare, inability to answer questions, and no subsequent recollection of events that occurred during the seizure. The attack usually lasts from 3 to 10 seconds and ends as abruptly as it began. It is a relatively rare form of epileptic seizure. However, the clinical pattern is so characteristic that, according to Temkin (1971), it was already recognized by Poupart in 1705. Other excellent descriptions may be found in the medical literature of the eighteenth (Tissot, 1770) and nineteenth centuries: "I have visited a young lady, whose father is epileptic. She had attacks during meetings, and when walking or riding. She does not fall down, her eyes are convulsive, with a fixed stare. The attack lasts no longer than a few seconds and the patient takes up the conversation at the point she left it, without suspecting anything that happened to her." (Esquirol, 1838).

A historical summary may help to understand why, until recent years, a variety of conditions were included under the heading of absence epilepsies. Absence epilepsies are electroclinical syndromes in which AS are associated with bilateral, symmetrical, synchronous, spike-and-slow-wave bursts at 3–4 Hz, but this striking EEG pattern was only discovered in 1935 (Gibbs et al., 1935).

Before 1935, diagnosis was made solely on clinical grounds. Many doctors oversimplified the classification of epilepsies: Grand Mal vs Petit Mal (Esquirol, 1838), epilepsia gravior (convulsive seizures) vs epilepsia minor (Reynolds, 1861; Gowers, 1881). For this reason, AS and partial complex seizures were not clearly delineated. Twenty years ago, the term "fausses absences temporales" was commonly used. All losses of awareness and responsiveness are not AS. They may be partial seizures of frontal or temporal origin.

After the discovery of a characteristic EEG pattern, the situation was clear . . . for a moment. As a matter of fact, it was a period of great electro-encephalographic enthusiasm. Gibbs et al. (1943) wrote, "It is more accurate to apply to a particular electroencephalographic pattern the name of the clinical type of seizures with which it is associated than to use purely descriptive terms . . . There can be no question of the propriety of speaking of a petit mal type of dysrhythmia when it occurs during a clinical petit mal seizure . . . We believe that it is equally proper to speak of a petit mal type of dysrhythmia when the same pattern appears in routine records in the absence of clinically obvious seizures, even if the particular patient has no history of petit mal or of epilepsy". This too broad conception was responsible for a tremendous amount of EEG misinterpretation. Epilepsy is a clinical phenomenon and not an EEG one. The issue was confused by including other rhythms morphologically fairly similar, but basically quite unrelated to absence epilepsies. Bilateral, symmetric and almost synchronous or more or less rhythmic spike-and-slow-wave complexes were found in patients with brain tumours. Misinterpretations of records also occurred in experimental research.

Again, because of the attractive accuracy of electroencephalography, EEG-based classifications of epilepsies were proposed. Under the heading of Petit Mal Triad, Lennox (1945) gathered diverse clinical manifestations sharing an EEG pattern of bilateral spike-waves. The triad included absences, myoclonic seizures and akinetic seizures. It was misunderstood and misused. Later, typical absences, or absence seizures, were opposed to atypical absences with a slower rhythm of spike-wave bursts, called Petit Mal Variant (Commission, 1981). The application of loose criteria, for clinical events as well as for EEG patterns, explains the wide discrepancies observed in the published data on petit mal or on absences.

However, the international classification of epileptic seizures described several types of AS: *simple absences*, with only an impairment of consciousness, *absences with mild clonic components*, restricted to the eyelids, or with twitching of the face or jerks of the head and the shoulders, *absences with atonic components*, resulting in a gradual lowering of the head and/or arms, but rarely causing the patient to fall, *absences with tonic components*, the eyes rotating upwards, less often the head or the trunk being drawn backwards, *absences with automatisms* and *absences with autonomic components*. This is probably an unnecessary complication, because in many patients several components are noted during a given AS, and/or they may present several types of AS (Penry et al., 1975). The duration of the attack is responsible for its variable features. The patient's age also plays a role. The practical value of identifying only individual seizure types is limited. They have to be placed in their context.

It is mandatory to stick to the International Classification of Epilepsies (Commission, 1989). This classification is based on the description of syndromes. An epileptic syndrome is an epileptic disorder characterized by a cluster of signs and symptoms customarily occurring together; these include such items as the type of seizure, aetiology, anatomy, precipitating factors, age of onset, severity, chronicity, diurnal and circadian cycling and prognosis. Two divisions shape the major classes of epilepsies. The first separates generalized epilepsies from focal epilepsies. The other distinction is an aetiological one: symptomatic epilepsies are the consequence of a known disorder of the central nervous system; the phenotypes of crypto-genic epilepsies are very close to those of symptomatic epilepsies, some of them are probably symptomatic but remain without demonstrable aetiology; and idiopathic epilepsies are defined by age-related onset, clear clinical and EEG characteristics and a patent or hidden genetic aetiology. In idiopathic epilepsies, acquired, environmental factors are likely to play a role in the clinical expression of a genetic propensity to epilepsy, but are of minor importance. All seizures are initially generalized with an EEG expression characterized by bilateral, synchronous, symmetrical discharges of spike-waves or polyspike-waves. The patient usually has a normal interictal state, without neurological or neuroradiological signs. Interictal EEGs show normal background activity and generalized discharges such as spikes, polyspikes, spike-waves or polyspike-waves.

All epilepsies with AS are idiopathic generalized epilepsies, mainly differing in age of onset and the place of AS in the clinical picture. AS are the sole or prominent feature in two epileptic syndromes and are associated with other types of seizures in a few other epileptic syndromes.

Childhood absence epilepsy

Childhood absence epilepsy is characterized as follows:

— AS are the initial type of attacks. Loss of consciousness is usually complete.
— The onset is between 2 and 12 years of age, with a peak at 6.7 years of age, in previously normal children.
— AS are frequent throughout the day — from 20 to 200 per day — occurring at random or precipitated by environmental factors.
— AS are concomitant with bilateral, synchronous and symmetrical spike-and-slow-wave bursts, at 3 Hz. They start and end abruptly, on a normal background activity.
— AS tend to vanish spontaneously and progressively during adolescence, and in adulthood very few patients continue to present with AS as the single type of seizure. AS are controlled in 80% of cases by specific anti-epileptic drugs, such as ethosuximide and valproate, alone or in combination.
— A durable remission is frequent, but it is held that 40% of patients develop, during adolescence or in the first part of adult life, generalized tonic-clonic seizures, usually infrequent and easily controlled.

P. Loiseau

Juvenile absence epilepsy

. Juvenile absence epilepsy is characterized as follows:

— The onset is around puberty, i.e. between 10 and 17 years of age.
— The other main characteristic is a low frequency of AS, not occurring every day, and mainly in the awakening period.
— AS are very similar to those of childhood absence epilepsy.

However, retropulsive components are less common and the disturbance of consciousness is less severe. The older the patient, the lighter the impairment of consciousness.

— On the ictal EEG, the spike-wave frequency is usually faster than 3 Hz: 3.5 to 4 Hz.
— The majority of patients — about 80% — also experience generalized tonic-clonic seizures, beginning before AS in about 50% of the patients. Most of these occur on awakening.
— AS are easily controlled. Usually generalized tonic-clonic seizures respond well to therapy, but relapses are frequent after stopping medication, even after a long seizure-free period.

Myoclonic juvenile epilepsy

This is a syndrome of idiopathic generalized epilepsy with an onset between 12 and 18 years of age in 80%. The characterizing feature of the syndrome is bilateral, arhythmic, irregular myoclonic jerks, predominantly of the arms, occurring shortly after awakening and often precipitated by sleep withdrawal. Sooner or later, generalized tonic-clonic seizures appear in more than 90% of the patients. AS are noted in 15 to 38% of the cases (Janz, 1985; Panayiotopoulos et al., 1989). They may appear several years before the first myoclonic jerk and, for this reason, are misdiagnosed. However, a slightly different EEG pattern could avoid a misdiagnosis (Panayiotopoulos et al., 1989).

Absence seizures in other idiopathic generalized epilepsies

An obvious overlap exists between the clear-cut phenotypes above. Rare AS may be noted — or, more often, recorded on the EEG — in epilepsy with grand mal seizures on awakening and in other idiopathic generalized epilepsies. In some children, generalized tonic-clonic seizures begin before 5 years of age, mainly in boys, and AS occur later (Doose et al., 1965; Dieterich et al., 1985a,b). They are more severe forms of epilepsy. Photosensitive epilepsies may present with AS (Jeavons et al., 1986). AS may also occur in brain-damaged children (Loiseau, 1985), owing to a genetic predisposition. Even if their natural history is that of an absence epilepsy, they cannot, of course, be diagnosed as childhood absence epilepsy.

Three main diagnostic problems exist:

— Atypical absences in cryptogenic or symptomatic generalized epilepsies. They belong to very different epileptic syndromes and are not absence epilepsies.
— Patients with features of both idiopathic and symptomatic epilepsies, such as in epilepsy with myoclonic absences (Tassinari and Bureau, 1984) or in Intermediate Petit Mal (Lugaresi et al., 1973). Even if they illustrate a biological continuum (Berkovic et al., 1987), they probably are not very useful for understanding the pathophysiology of AS (Mirsky et al., 1986).
— AS in patients with frontal lesions. These pose an important and difficult problem. At least for a time, medial frontal epilepsies may present with AS. A search for clinical or EEG peculiarities is often disappointing and only the outcome permits a correct diagnosis.

In conclusion: "All classifications in all sciences make distinctions more exact and abrupt than any that exists in nature." (Jackson, 1958).

References

Berkovic SF, Andermann F, Andermann E, Gloor P (1987) Concepts of absence epilepsies: discrete syndromes or biological continuum? Neurology 37: 993–1000

Commission on Classification and Terminology of the International League Against Epilepsy (1981) Proposal for revised clinical and electroencephalographic classification of epileptic seizures. Epilepsia 22: 489–501

Commission on Classification and Terminology of the International League Against Epilepsy (1989) Proposal for classification of epilepsies and epileptic syndromes. Epilepsia 30: 389–399

Dieterich E, Baier WK, Doose H, Tuxhorn I (1985a) Longterm follow-up of childhood epilepsy with absences. I. Epilepsy with absences at onset. Neuropediatrics 16: 149–154

Dieterich E, Doose H, Baier WK, Fichsel H (1985b) Longterm follow-up of childhood epilepsy with absences. II. Absence-epilepsy with initial grand mal. Neuropediatrics 16: 155–158

Doose H, Völzke E, Scheffner D (1965) Verlaufsformen kindlicher Epilepsien mit Spike wave-Absenzen. Arch Psychiatr Z Neurol 207: 394–415

Esquirol J (1838) De l'epilepsie. In: Traité des Maladies Mentales. Tome 1. Baillére, Paris, pp 274–335

Gibbs FA, Davis H, Lennox WG (1935) The EEG in epilepsy and in conditions of impaired consciousness. Arch Neurol 34: 1134–1148

Gibbs FA, Gibbs EL, Lennox WG (1943) Electroencephalographic classification of epileptic patients and control subjects. Arch Neurol 50: 111–128

Gowers WR (1881) Epilepsy and other chronic convulsive diseases: their causes, symptomes and treatment. Churchill, London

Jackson JH (1958) Selected writings. In: Taylor J (ed) Staples Press, London, p 202

Janz D (1985) Epilepsy with impulsive petit mal (Juvenile Myoclonic Epilepsy). Acta Neurol Scand 72: 449–459

Jeavons PM, Bishop A, Harding GFA (1986) The prognosis of photosensitivity. Epilepsia 27: 569–573

Lennox WG (1945) The Petit Mal epilepsies. J Am Med Assoc 129: 1069–1073

Loiseau P (1985) Childhood absence epilepsy. In: Roger J, Dravet C, Bureau M, Dreifuss F, Wolf P (eds) Epileptic syndromes in infancy, childhood and adolescence. Libbey Eurotext, London, pp 106–120

Lugaresi E, Pazzaglia P, Franck L, Roger J, Bureau-Paillas M, Ambrosetto G, Tassinari CA (1973) Evolution and prognosis of primary generalized epilepsies of the petit mal absence type. In: Lugaresi E, Pazzaglia P, Tassinari CA (eds) Evolution and prognosis of epilepsies. Aulo Gaggi, Bologna, pp 1–22

Mirsky AF, Duncan CC, Myslobodsky MS (1986) Petit Mal Epilepsy: a review and integration of recent information. J Clin Neurophysiol 3: 179–208

Panayiotopoulos CP, Obeid T, Waheed G (1989) Differentiation of typical absence seizures in epileptic syndromes. A video EEG study of 224 seizures in 20 patients. Brain 112: 1039–1056

Penry JK, Porter RJ, Dreifuss FEI (1975) Simultaneous recording of absence seizures with video tape and electroencephalography. A study of 374 seizures in 48 patients. Brain 98: 427–440

Poupart (1945) In: Temkin O (ed) The falling sickness. Johns Hopkins Press, Baltimore, p 250

Reynolds JR (1861) Epilepsy: its symptoms, treatment, and relation to other chronic convulsive diseases. Churchill, London

Tassinari CA, Bureau M (1984) Epilepsies avec absences myocloniques. In: Roger J, Dravet C, Bureau M, Dreifuss F, Wolf P (eds) Les syndromes épileptiques de l'enfant et de l'adolescent. Libbey Eurotext, Paris, pp 123–131

Temkin O (1971) The falling sickness, 2nd ed. Johns Hopkins Press, Baltimore

Tissot DM (1770) Traité de l'Epilepsie, faisant tome troisième du Traité des Nerfs et de leurs Maladies. Didot, Paris

Author's address: Dr. P. Loiseau, Service de Neurologie, C.H.U., Hôpital Pellegrin, Place Amélie Raba-Léon, F-33076 Bordeaux Cedex, France

J Neural Transm (1992) [Suppl] 35: 7–19

Pharmacological models of generalized absence seizures in rodents

O. Carter Snead III

Division of Neurology, Children's Hospital, Los Angeles, and Department
of Neurology, University of Southern California School of Medicine,
Los Angeles, California, U.S.A.

Summary. A number of animal models of generalized absence seizures in rodents are described. These include absence seizures induced by γ-hydroxybutyrate (GHB), low dose pentylenetetrazole, penicillin, THIP, and AY-9944. All of these models share behavioral and EEG similarity to human absence seizures and show pharmacologic specificity for antiabsence drugs such as ethosuximide and trimethadione. Moreover, the absence seizures induced by these agents are exacerbated by GABAergic agonists, a property unique to experimental absence seizures. These models are predictable, reproducible, and easy to standardize. They are useful both in studying mechanisms of pathogenesis of absence seizures as well as in screening for antiabsence activity of potential antiepileptic drugs.

Generalized absence seizures are fundamentally different from any other kind of human epilepsy (Berkovic et al., 1987). Pharmacologic models of absence in any species should reflect this difference to accurately mimic the human condition. Clinically, generalized absence seizures are characterized by a number of striking features (Table 1). The hallmark of generalized absence seizures is the occurrence of 3 per second, bilaterally synchronous spike and slow wave discharges (SWD). The behavior associated with this electrographic discharge consists of immobility, starring, occasional automatisms, and occasionally some loss of body tone with a myoclonic component (Penry et al., 1975; Lockman, 1989). Another distinguishing feature of generalized absence seizures is the fact that these occur almost exclusively in children between the ages of 4 and 15. Therefore, this is a disorder of developing brain.

Generalized absence seizures are pharmacologically unique clinically. The antiepileptic drugs ethosuximide and trimethadione, are effective in generalized absence seizures, but not in other kinds of seizures. Valproate, which is also effective in generalized absence seizures, is also effective in other kinds of seizure disorders, but is biochemically unique from all other anticonvulsants. Antiepileptic drugs which are effective in generalized convulsive and partial epilepsies, i.e. phenytoin and carbamazepine, are

Table 1. Characteristics of generalized absence seizures in human

1. Occur in children with onset between 4–15 years of age
2. EEG findings of bilaterally synchronous 2.5–3 c/s SWD
3. Clinical findings of staring, behavioral arrest, occasional myoclonus, eye movements, automatisms
4. Brief, no aura, no postictal state
5. Treated with ethosuximide, trimethadione, or valproic acid
6. Aggravated by carbamazepine and phenytoin
7. Aggravated by GABA agonist Progabide

known to make generalized absence seizures worse (Roseman, 1961; Snead and Hosey, 1985).

Effective and valid animal models for generalized absence seizures should incorporate as much as possible all of these distinguishing features of the clinical condition. In addition, there are several other criteria, based on available data concerning the pathophysiology of absence seizures which animal models of generalized absence seizures should satisfy.

The precise mechanism of production of bilaterally synchronous SWDs in generalized absence seizures is not known; however, there is a large body of literature which points to involvement of thalamocortical mechanisms in the pathogenesis of generalized absence seizures. Data from both human (Williams, 1953) and animal studies point to the thalamocortical loop as the primary abnormality in generalized absence epilepsy. For example, studies in the feline penicillin model of generalized absence epilepsy model suggest that the intramuscular penicillin used to induce SWD results in diffuse hyperexcitability of cortical neurons. Normally, these cortical neurons respond to thalamocortical volleys with spindles, but in this hyperexcitable state, SWD are the result of thalamocortical volleys. The thalamus, too, is involved in this process. It rapidly becomes actively involved and constitutes an essential component of the neuronal system that sustains the SWD.

Generalized SWD thus reflect a wide-spread, phase locked, oscillation between excitation spike and inhibition wave in mutually interconnected thalamocortical neuronal networks. Both the cortex and thalamus are essential for the maintenance of the spike wave rhythm (Gloor, 1984).

Other subcortical structures such as the mamillary bodies and their projections (Mirski et al., 1986; Mirski and Ferendelli, 1986), the superior colliculus (Depaulis et al., 1990), and the substania nigra (Depaulis et al., 1988, 1989) may also have an important role in generalized absence seizures experimentally, but their involvement in human absence has yet to be established. Thus, an additional criteria for animal models of absence should be involvement of thalamocortical mechanisms in the genesis of the SWD.

A final requirement of generalized absence seizure models, is that they be exacerbated by both direct and indirect GABA agonists. GABAergic agonists, both direct (muscimol, THIP) and indirect (uptake inhibitors and

Table 2. Criteria for experimental generalized absence seizures

1. EEG and behavior similar to the human condition
2. Reproducibility and predictability
3. Quantifiable
4. Appropriate pharmacology
5. Unique developmental profile
6. Exacerbated by GABAergic drugs
7. Involvement of thalamocortical mechanisms

GABA-T inhibitors), are generally anticonvulsant in animal models of generalized convulsive or partial epilepsies (Snead, 1983). However, in a number of animal models of generalized absence seizures including the pentylenetetrazol model, the genetic model of spontaneous SWD in Wistar rat, the γ-hydroxybutyrate model, and the WAG-Rij strain of SWD rat, GABAergic agonists are proconvulsant (Mysloblodsky et al., 1979; Vergnes et al., 1984; Snead, 1984, 1990; Peeters, 1989). There is a clinical corollary to this phenomenon, since vigabatrin, a GABA-T inhibitor, and progabide, a GABA agonist have been shown to exacerbate generalized absence seizures in humans (Loscher, 1982; Van der Linden et al., 1981). The mechanism of this proconvulsive effect of GABA on experimental and clinical generalized absence seizures is unknown. The hypothesis that generalized absence seizures are a manifestation of inhibitory processes and are secondary to a paroxysmal activity in cortical inhibitory pathways was developed by Fromm and Kohli (1972) and more recently by Gloor and Fariello (1988). The evidence for this premise is based on the exacerbation of clinical and experimental absence by GABA agonists and the observation that in the feline generalized penicillin model of absence seizures, there is no refractoriness to the hyperpolarizing effect of GABA during SWD. Rather there is preservation of the inhibitory postsynaptic potential in this model (Kostopoulos, 1986).

Based on the foregoing observations, the following are proposed criteria for generalized absence seizure synthesis (Table 2): A. EEG and behavioral similarities with the human condition; B. easy reproducibility; C. predictable development and course; D. quantifiability; E. pharmacologic specificity for anti-absence drugs such as ethosuximide; F. distinct developmental profile; that is, the seizures should be most prominent in developing animals; G. potentiation of the model by direct and indirect GABA agonists; H. involvement of thalamocortical mechanisms in the generation of the bilaterally synchronous SWD.

The experimental models discussed in this paper are pharmacologic models in rodents which meet all or some of these criteria. This discussion is limited to rodent models because these animals are easily obtainable, and easier to work with than larger animals such as cats and monkeys. When using rodent models of experimental absence seizures however, one must bear in mind the work of McQueen and Woodbury (1975), who attempted to produce bilaterally synchronous SWD in the electrocortico-gram of rats by a number of experimental paradigms. These included administration of pentylentetrozole, picrotoxin, conjugated estrogens, and bilateral intracerebral cobalt implants. No pharmacologic modality produced consistent, bilaterally synchronous SWD. The authors concluded, therefore, that the rodent was not suitable for any detailed study of the pathophysiology of SWD. Since that work was published, however, a number of rat models have been developed which have direct relevance to generalized absence seizures. Although rodents apparently are indeed incapable of generating 3 per second SWD, generalized absence seizures may, in fact, be defined in this species by pharmacologic, developmental, and behavioral characteristics described above. Even though rodents lack the thalamocortical circuitry to generate 3/sec SWD, recent data in both genetic and pharmacologic models of absence indicate that thalamocortical mechanisms are involved in the genesis of bilaterally synchronous SWD in this species (Vergnes et al., 1987, 1989).

The pharmacologic models of generalized absence seizures to be dis-cussed include SWD induced by γ-hydroxybutyrate, pentylenetetrazol, penicillin, THIP, and AY9944. These are all electrographic models since bilaterally synchronous SWD should be the sine qua non for experimental absence.

The GHB model

γ-Hydroxybutyrate (GHB) is a γ-amniobutyric acid (GABA) metabolite which occurs naturally in mammalian brain (Roth and Giarman, 1969). When given to animals, GHB produces a predictable sequence of electro-graphic and behavioral events which closely resemble generalized absence seizures. This phenomena has been well described in cats, rats, and monkeys (Snead, 1976, 1978; Godschalk et al., 1977; Bearden et al., 1980; Snead et al., 1980).

In order to enhance the reproducibility and predictability of the GHB model of generalized absence seizures, we have chosen to use the pro-drug of GHB, γ-butyrolactone (GBL) to induce SWD. This drug is used because of consistency and rapidity of onset of its effect (Bearden et al., 1980). It has been shown to produce exactly the same EEG and behavioral effect as that of GHB (Snead et al., 1980). However, recently, there has been concern that GBL itself may have some intrinsic biological activity (Vayer et al., 1987). If true, this would negate the validity of the use of GBL as a pro-drug for GHB in the induction of SWD. In view of this concern the hypothesis that the epileptogenic effects of GBL are due solely to its conversion to GHB has been tested (Snead, 1991). The regional brain concentration of both GHB and GBL was determined in time course and dose response studies after i.p. administration of GBL, as well as at the onset of EEG changes induced by both GHB and GBL. Also, the EEG

and behavioral effects of both drugs were ascertained after bilateral intrathalamic microinjection in the rat.

GBL produced a rapid onset of bilaterally synchronous SWD in rat
which correlated with a rapid appearance of GHB in brain. In the GHB-
treated animals, EEG changes occurred 20 minutes after GHB administration when GHB levels in brain were peaking. The threshold brain
concentration of GHB for EEG changes in both GHB and GBL-treated
animals was 240 μM. GBL concentration in brain peaked one minute after
GBL administration and fell rapidly to undetectable levels within five
minutes. Bilateral microinjection of GHB into thalamus resulted in a brief
burst of SWD, while GBL administered into the thalamus had no effect.
These data demonstrate the validity of the use of GBL as a pro-drug for
GHB in this model of generalized absence seizures.

Based on these data, the GHB model has now been standardized and
is used routinely in our laboratory. A standard dose of 0.13 ml (150 milligrams) GBL per kilogram given intraperitoneally (i.p.) reliably produces
onset of bilaterally synchronous SWD within 2–5 minutes of GBL administration. The frequency of the SWD is 7–9 cycles per second. Associated
with these hypersynchronous electrographic changes are behavioral arrest,
facial myoclonus, and vibrissal twitching. Therefore, this model meets the
first two criteria outlined in Table 1 in that it is predictable, reproducible,
and produces electrographic and behavioral events similar to the human
condition. An additional advantage of the GHB model is that it affords

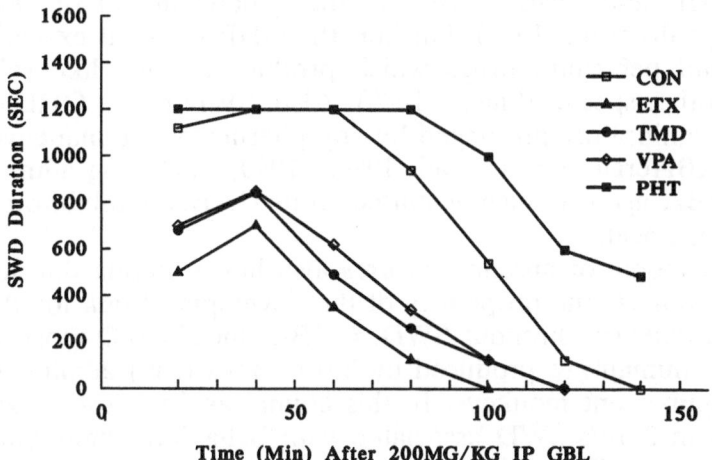

Fig. 1. Duration of GHB-induced SWD in a series of animals treated with various
anticonvulsants. The dose of GBL was 150 mg/kg. Each data point represents the mean
of 4 animals. The standard error was 15% or less in all experiments and is not shown.
All antiabsence drugs resulted in a significant (p < 0.05; Wilcoxon) reduction of SWD
duration while phenytoin significantly enhanced SWD duration. The dose of drugs was
ethosuximide (ETX): 100 mg/kg; trimethadione (TMD): 150 mg/kg; valproic acid
(VPA): 200 mg/kg; and phenytoin (PHT) 30 mg/kg. Each drug was given 60 min prior
to GBL

control of pharmacokinetic variables in any pharmacologic study since the concentration of GBL and GHB can be determined in brain and the kinetics are known.

The GHB model of generalized absence seizures is quantitated in a matter similar to other electrographic models of generalized absence seizures (Depaulis et al., 1989). GHB-induced SWD may be quantitated in terms of cumulative duration (sec) per 20 minute epoch of time (Fig. 1) or as a percent of control SWD duration. In this way, the GHB model of absence may be compared with any other rodent model of generalized absence seizures using the same pharmacologic paradigm (Depaulis et al., 1989).

As mentioned above, the rat is not capable of generating 3 per second spike and slow wave discharge. However, the frequency and morphology of SWD observed in the GHB model of absence seizures are quite similar to those produced by low dose pentylenetetrazole (PTZ) and THIP, two other pharmacologic models of absence in rodent (Depaulis et al., 1989; Fariello and Golden, 1987; Marescaux et al., 1984), as well as the SWD seen in, a genetic model of absence, spontaneous SWD in Wistar rat (Marescaux et al., 1984).

The pharmacology of the GHB model of generalized absence seizures is what one would predict for absence seizures (Godschalk et al., 1976; Snead, 1978a,b). GHB induced SWD are significantly decreased by anti-absence drugs such as ethosuximide, trimethadione, and valproate and enhanced by phenytoin (Fig. 1). Another aspect of the pharmacology of this model is that both GHB and GBL exacerbate the genetic model of generalized absence (Depaulis et al., 1988). Further, the GHB model is exacerbated by both PTZ and penicillin, drugs which produce absence-like seizures in rodents and other species (Snead, 1988). Also, SWD in the GHB model of absence are significantly prolonged by any pharmacologic maneuver which enhances GABAergic tone (Snead, 1984a, 1990), including administration of direct GABA agonists such as muscimol or GABA-T inhibitors, such as amino-oxyacetic acid.

The GHB model of absence seizures also has a unique developmental profile which reflects the propensity of the developing brain for the occurrence of bilaterally synchronous SWD. GHB-induced SWD most similar to those seen in human are produced by intravenous (i.v.) administration of GHB to prepubescent monkeys. In this animal an i.v. dose of 200 mg/kg GHB results in 2.5 c/s SWD associated with behavioral immobility, head drops, starring, pupillary dilation, eyelid fluttering, rhythmic eye movements, and stereotyped automatisms (Snead, 1978a).

In rodent, the age at which animals are most sensitive to GHB is during the fourth postnatal week of life (Snead, 1984b). At this same time point in development, the rat is also most sensitive to enhancement of GHB-induced SWD by GABAergic agonists (Snead, 1990).

There is evidence that the thalamus is involved in the GHB model of absence in rodents. Vergnes and others (Vergnes et al., 1987) have demonstrated using mapping of EEG with bipolar depth recordings that SWD originate in the thalamus and cortex in this model and other rodent models of absence seizures. The hippocampus was not involved in any of these rodent models of absence seizures. In addition, Vergnes et al. (this volume) have shown that lesioning the lateral thalamus in rats in whom the corpus callosum has been cut to disrupt bilateral synchrony, results in blockade of SWD over the lesioned side in animals given GHB or PTZ or in Wistar rats with spontaneous SWD. Finally, bilateral microinjection of GHB into the lateral thalamus of rat results in SWD (Snead, 1991). When GHB is given in this fashion to Wistar rats with spontaneous SWD, absence status epilepticus results (Liu et al., 1991).

In conclusion, the GHB rat model of generalized absence seizures meets all criteria which have been put forth to date for experimental absence seizures. This is a useful experimental model for the study of the mechanisms of bilaterally synchronous SWD production, and can be used to screen for anti-absence activity of potential antiepileptic drugs.

The PTZ model

Drugs effective in the treatment of absence seizures have been reported to be more potent against clonic seizures induced by PTZ in a dose of 40 milligrams per kilogram or higher than against seizures produced by maximal electroshock. A stronger relative potency against PTZ seizures has thus been used to predict efficacy against absence seizures in drug screening programs (Swinyard, 1969; Krall et al., 1978). However, traditionally the PTZ model has been pharmacologic rather than electrographic because bilaterally synchronous SWD produced by low doses of this drug were not reported in rodents until relatively recently when Marescaux and others reported the utility of using lower doses of 10 and 20 mg/kg PTZ to produce electrographic events more similar to generalized absence seizures, that is bilaterally synchronous SWD associated with behavioral arrest and immobility (Marescaux et al., 1984). PTZ induced seizures are predominantly dose-dependent with low dose (10–20 mg/kg) producing the SWD model described below, an intermediate dose (40–60 mg/kg) producing spike trains with clonic seizures responsive to anti-absence drugs and high dose (≥80 mg/kg) producing tonic seizures unresponsive to anti-absence drugs. The idea that the clonic component of PTZ seizures are more representative of absence seizures than the tonic components finds support in the work of Browning and Nelson (1986) who have shown that tonic and some types of clonic seizures occur independently of seizure discharge in forebrain structures whereas clonus restricted to the face and forelimbs depends on seizure discharge emanating from structures within the forebrain for expression. Facial clonus is typical of all rat models of generalized absence seizures.

The low dose PTZ model would seem to meet all the criteria set forth in Table 2. A dose of 20 milligrams per kilogram of PTZ results in bursts of bilaterally synchronous SWD with a frequency of approximately 7–9 c/s. The behavior seen in the PTZ animal is exactly the same as that described for GHB-treated animals, as well as during spontaneous, SWD in the genetic model in Wistar rat. It should be reiterated here that the facial clonus is representative of forebrain involvement.

Marescaux et al. have shown that the PTZ model meets the pharmacologic specificity requirement for experimental absence seizures; i.e. it is aborted by the use of anti-absence drugs such as ethosuximide and made worse by phenytoin and carbamazepine (Marescaux et al., 1984). PTZ seizures also have a distinct ontogeny with spikes appearing at the 12th post natal day in rats and the wave form at a later date (Zouhar et al., 1980; Mares et al., 1982). PTZ-induced seizures are potentiated both by GABAergic agonists (Mysloblodsky, 1986) as well as by GHB and muscimol (Unnerstall and Pizzi, 1981). The dose of PTZ to induce SWD is 20 milligrams per kilogram of PTZ given intraperitoneally in a volume of 1 cc per kilogram. The same techniques as described for the GHB model are used in terms of the quantitation of the model by measuring SWD duration per 20 minute epochs (Depaulis et al., 1989). This is a useful, predictable, reproducible, model of generalized absence seizures which can be used along with other pharmacologic or genetic models in rodents to investigate basic mechanisms of SWD production as well as to screen drugs for anti-absence activity.

The penicillin model

Intramuscular administration of penicillin to the cat consistently produces generalized, bilaterally synchronous SWD discharges associated with blinking, myoclonus, and starring (Gloor, 1984; Fisher and Prince, 1977; Taylor-Courval and Gloor, 1984; Guberman et al., 1975). Furthermore, in the cat, this model shows pharmacological specificity for anti-absence drugs (Guberman et al., 1975) and is exacerbated by PTZ (Gloor and Testa, 1974), photic stimulation (Quesney, 1984), and GABAergic agonists (Fariello, 1979). In fact, the feline generalized penicillin model of generalized absence seizures with bilaterally synchronous SWD is the best characterized and most carefully studied of all animal models of generalized absence seizures to date (Gloor, 1984).

When given intramuscularly to rodents, penicillin does not consistently produce bilaterally synchronous SWD similar to that seen in cats. Rather this drug produces multifocal spikes with only occasional bursts of bilaterally synchronous SWD associated with a decrease in vigilance (Avoli, 1980). The penicillin model in rodents has not been as well characterized as the GHB and PTZ model and is of limited usefulness because of inconstant penetration of penicillin into the brain through the blood brain barrier. This model, however, has been shown to be exacerbated by GHB. Conversely, pretreatment with penicillin prolongs GHB induced SWD (Snead, 1988). When using this model of absence, the same general experimental design as described above is used. The dose of penicillin is from 300,000–600,000 units per kilogram given intramuscularly. There are no anti-absence, antiepileptic drug or ontogeny data for the penicillin model in rat, but there is some evidence of involvement of thalamocortical mechanisms in penicillin-induced SWD in rat (Avoli, 1980). A possible limitation in the usefulness of this model is the development of tolerance. Bo et al. (1984) have reported a significant reduction in "spike frequency", in rats receiving, 1,000,000 μ/kg, a very high dose, every 48 hours.

The THIP model

Recently, Fariello and Golden have purposed the use of THIP [4,5,6,7 tetrahydroxyisoxazolo (4,5,c) pyridine 3-ol], a GABA agonist, to induce bilaterally synchronous SWD in rats (Fariello and Golden, 1987). This compound when given intraperitoneally in a dose of 5–10 milligrams per kilogram, results in bilaterally synchronous SWD which occur in bursts lasting 1 to 7 seconds. The frequency of the discharge is approximately 4–6 cycles per second. Behaviorally, the animals have immobility and some vibrissal twitching. This model appears to be a generalized absence model based on electroclinical correlations; however, to date there are no pharmacologic data to support this, nor are there ontogeny data, or data concerning possible thalamocortical mechanisms in the generation of these discharges. This model is different from other described in this article because it is exacerbated by valproate (Vergnes et al., 1985). The dose of THIP to elicit SWD is 7.5 milligrams per kilogram of THIP given i.p. in a volume of normal saline of 1cc per kilogram. The model is quantitated in the same manner as described above for the GHB model (Depaulis et al., 1989).

AY-9944 model

Recently the drug AY-9944, a compound which inhibits the reduction of 7-dehydrocholesterol to cholesterol (Cenedella, 1980; Dvornik and Hill, 1968), has been reported to produce absence like seizures in rodents (Smith and Bierkamper, 1990). AY-9944-treated rats averaged about 50 generalized absence seizures per hour lasting 2–15 seconds. These episodes reportedly had the same electrographic and behavioral characteristics as those described for the above pharmacologic models as well as the genetic models reported elsewhere in this volume. These seizures seem to be specifically reduced by anti-absence drugs and exacerbated by phenytoin and GABA agonists (Smith and Bierkamper, 1990). There are no ontogeny data available for this model. Male Long Evans rats are given AY 9944 i.p. in a dose of 7.5 mg/kg in olive oil (10 ml/kg) beginning at 2 days of post natal age. Several subsequent injections are given once every 6 days until day fifty. This seems to induce a permanent seizure state.

This has been a brief, descriptive discussion of experimental absence seizures in the rat. Rodent models of absence, both pharmacologic and genetic, have a useful place in the schema of experimental epilepsy in spite of species differences in the neurophysiology and synaptic circuitry of thalamocortical systems (Steriade and Llinas, 1988). These models can all be quantitated in the same fashion (Depaulis et al., 1989), and can be used in the same experimental paradigm, whether it be designed to investigate mechanisms underlying SWD production in absence or to screen for antiabsence activity in potential antiepileptic drugs.

Acknowledgment

This work was supported in part by Grant No. NS 17117 from the NINDS.

References

Avoli M (1980) Electroencephalographic and pathophysiologic features of rat parenteral penicillin epilepsy. Exp Neurol 69: 373–382

Bearden LJ, Snead OC, Healy CT, Pegram GV (1980) Antagonism of gamma hydroxybutyrate-induced frequency shifts in the cortical EEG of rats by dipropylacetate. Electroencephalogr Clin Neurophysiol 49: 181–183

Berkovic SF, Andermann F, Andermann E, Gloor P (1987) Concepts of absence epilepsies: discrete syndromes or biological continuum? Neurology 37: 993–1000

Black JA, Golden GT, Fariello RG (1980) Ketamine activation of experimental corticoreticular epilepsy. Neurology 30: 315–318

Bo GP, Fonzari M, Scotto PA, Benassi E (1984) Parenteral penicillin epilepsy: tolerance to subsequent treatments. Exp Neurol 85: 229–232

Browning RA, Nelson DK (1986) Modification of electroshock and pentylenetetrazol seizure patterns in rats after pericollicular transections. Exp Neurol 93: 546–556

Cenedella RJ (1980) Concentration dependent effects of AY-9944 and U18-666A on sterol synthesis in brain. Variable sensitivities of metabolic steps. Biochem Pharmacol 29: 2753–2760

Depaulis A, Bourguignon J, Marescaux C, Vergnes M, Schmitt M, Micheletti G, Warter JM (1988) Effect of γ-hydroxybutyrate and γ-butyrolactone derivatives on spontaneous generalized non-convulsive seizures in the rat. Neuropharmacology 27: 6863–689

Depaulis A, Vergnes M, Marescaux C, Lannes B, Warter JM (1988) Evidence that activation of GABA receptors in the substantia nigra suppresses spontaneous spike and wave discharges in the rat. Brain Res 448: 20–29

Depaulis A, Snead OC, Marescaux C, Vergnes M (1989) Suppressive effects of intranigral injection of mucsimol in three models of generalized non-convulsive epilepsy induced by chemical agents. Brain Res 1498-64-72

Depaulis A, Liu Z, Vergnes M, Marescaux C, Micheletti G, Warter JM (1990) Suppression of spontaneous generalized non-convulsive seizures in the rat by microinjection of GABA antagonists into the superior colliculus. Epilepsy Res 5: 192–198

Dvornik D, Hill P (1968) Effects of long term administration of AY-9944, an inhibitor of 7-dehydrocholesterol delta[7]-reductase on serum and tissue lipids in the rat. J Lipid Res 9: 587–592

Fariello RG (1979) Action of inhibitory amino acids on acute epileptic foci: an electrographic study. Exp Neurol 66: 55–63

Fariello RG, Golden GT (1987) The THIP induced model of bilateral synchronous spike and wave in rodents. Neuropharmacology 26: 161–165

Fisher RS, Prince DA (1977) Spike-wave rhythms in cat cortex induced by parenteral penicillin. II. Cellular features. Electroencephalogr Clin Neurophysiol 42: 625–639

Fromm GH, Kohli CM (1972) The role of inhibitory pathways in petit mal epilepsy. Neurology 22: 1012–20

Gloor P (1984) Electrophysiology of generalized epilepsy. In: Schwartzkroin PA, Wheal (eds) Electrophysiology of epilepsy. Academic Press, New York, pp 107–136

Gloor P, Testa G (1974) Generalized penicillin epilepsy in the cat: effects of intracarotid and intravertebral pentylenetetrazol and amobarbital injections. Electroencephalogr Clin Neurophysiol 36: 499–515

Gloor P, Fariello RG (1988) Generalized epilepsy: some of its cellular mechanisms differ from those of focal epilepsy. Trends Neurol Sci 11: 63–68

Godschalk M, Dzoljic MR, Bonta IL (1976) Antagonism of gamma-hydroxybutyrate-induced hypersynchronization in the ECoG of rat by anti-petit mal drugs. Neurosci Lett 3: 1173–1178

Godschalk M, Dzoljic MR, Bonta IL (1977) Slow wave sleep and a state resembling absence epilepsy induced in the rat by γ-hydroxybutyrate. Eur J Pharmacol 44: 105–111

Guberman A, Gloor P, Sherwin AL (1975) Response of generalized penicillin epilepsy in the cat to ethosuximide and diphenylhydantoin. Neurology 25: 758–764

Kent AP, Webster RA (1986) The role of GABA and excitatory amino acids in the development of the leptazol-induced epileptogenic EEG. Neuropharmcology 25: 1023–1030

Kostopoulos G (1986) Neuronal sensitivity to GABA and glutamate in generalized epilepsy with spike and wave discharges. Exp Neurol 92: 20–36

Krall RL, Penry JK, Whiite BG, Kupferberg JH, Swinyard EA (1978) Antiepileptic drug development. II. Anticonvulsant drug screening. Epilepsia 22: 107–122

Liu Z, Snead OC, Vergnes M, Depaulis A, Marescaux C (1991) Intrathalamic injections of gamma-hydroxybutyric acid increase genetic absence seizures in rats. Neurosci Lett 125: 19–21

Lockman LA (1989) Absence, myoclonic and atonic seizures. Pediatr Clin North Am 36: 331–343

Loscher W (1982) Comparative assay of anticonvulsant and toxic potencies of sixteen GABAmimetic drugs. Neuropharmacology 21: 803–810

Mareš P, Marešova D, Trojan S, Fischer J (1982) Ontogenetic development of rhythmic thalamocortical phenomena in the rat. Brain Res Bull 8: 765–769

Marescaux C, Micheletti G, Vergnes M, Depaulis A, Rumbach L, Warter JM (1984) A model of chronic spontaneous petit mal-like seizures in the rat: comparison with pentylenetetrazole-induced seizures. Epilepsia 25: 326–331

McQueen JK, Woodbury DM (1975) Attempts to produce spike-wave complexes in the electrocorticogram of the rat. Epilepsia 16: 295–299

Mirski MA, Ferendelli JA (1986) Anterior thalamic mediation of generalized pentylenetetrazol seizures. Brain Res 399: 212–223

Mirski MA, McKeon AC, Ferendelli JA (1986) Anterior thalamus and substantia nigra: two distinct structures mediating experimental generalized seizures. Brain Res 397: 377–380

Mysloblodsky MS, Ackerman RF, Engel J (1979) Effects of γ-acetylenic GABA and γ-vinyl GABA on metrazol-activated and kindled seizures. Pharmacol Biochem Behav 11: 265–271

Peeters BWMM, Van Rijn CM, Vossen JMH, Coenen AML (1989) Effects of GABA-ergic agents on spontaneous non-convulsive epilepsy, EEG and behavior, in the WAG/RIJ inbred strain of rats. Life Sci 45: 1171–1176

Penry JK, Porter RJ, Dreifuss FE (1975) Simultaneous recording of absence seizures with videotape and electroencephalography. A study of 374 seizures in 48 patients. Brain 98: 427–440

Quesney LF (1984) Pathophysiology of generalized photosensitive epilepsy in the cat. Epilepsia 25: 61–69

Roseman E (1961) Dilantin toxicity: a clinical and EEG study. Neurology 11: 912–917

Roth RH, Giarman NJ (1969) Conversion in vivo of γ-aminobutyric acid to γ-hydroxybutyrate in the rat. Biochem Pharmacol 18: 247–250

Smith KA, Bierkamper GG (1990) Paradoxical role of GABA in a chronic model of petit mal (absence)-like epilepsy in the rat. Eur J Pharmacol 176: 45–55

Snead OC (1978) Gamma-hydroxybutyrate in the monkey. I. Electroencephalographic, behavioral, and pharmacokinetic studies. Neurology 28: 636–642

Snead OC (1978) Gamma-hydroxybutyrate in the monkey. II. Effect of chronic oral anticonvulsant drugs. Neurology 28: 643–648

Snead OC (1978) Gamma-hydroxybutyrate in the monkey. III. Effect of intravenous anticonvulsant drugs. Neurology 28: 1173–1182

Snead OC (1983) On the sacred disease: the neurochemistry of epilepsy. Int Rev Neurobiol 24: 94–180

Snead OC (1984a) Gamma-hydroxybutyric acid, γ-aminobutyric acid and petit mal epilepsy. In: Fariello RG, Morselli PL, Lloyd KG, Quesney LF, Engel J (eds) Neurotransmitters, seizures, and epilepsy, II. Raven Press, New York, pp 37–47

Snead OC (1984b) Ontogency of γ-hydroxybutyric acid. II. Electroencephalographic effects. Dev Brain Res 15: 89–96

Snead OC (1988) γ-hydroxybutyrate model of generalized absence seizures: further characterization and comparison with other absence models. Epilepsia 29: 361–368

Snead OC (1990) The ontogeny of muscimol enhancement of the γ-hydroxybutyrate model of generalized absence seizures. Epilepsia 31: 363–368

Snead OC (1991) The γ-hydroxybutyrate model of absence seizures: correlation of regional brain levels of γ-hydroxybutyric acid and γ-butyrolactone with spike wave discharges. Neuropharmacology 30: 161–167

Snead OC, Hosey LC (1985) Exacerbation of seizures in children by carbamazepine. N Engl J Med 313: 916–921

Snead OC, Yu RK, Huttenlocher PR (1976) Gamma-hydroxybutyrate: correlation of serum and cerebrospinal fluid levels with electroencephalographic and behavioral effects. Neurology 26: 51–56

Snead OC, Bearden LJ, Healy CT, Pegram V (1980) Effect of acute and chronic anticonvulsant administration on endogenous γ-hydroxybutyrate in rat brain. Neuropharmacology 19: 47–52

Steriade M, Llinas RR (1988) The functional state of the thalamus and the associated neuronal interplay. Physiol Rev 68: 649–742

Swinyard EA (1969) Laboratory evaluation of antiepileptic drugs. Epilepsia 10: 107–119

Taylor-Courval D, Gloor P (1984) Behavioral alterations associated with generalized spike and wave discharges in the EEG of the cat. Exp Neurol 83: 167–186

Unnerstall JR, Pizzi WJ (1981) Muscimol and γ-hydroxybutyrate: similar interactions with convulsant agents. Life Sci 29: 337–344

Van der Linden GJ, Meinardi H, Meijer SWA, Bossi L, Gomeni CA (1981) A double blind crossover trial with progabide (SL76002) against placebo in patients with secondary generalized epilepsy. In: Dam M, Gram L, Penry JK (eds) Advances in Epileptology: XIIth Epilepsy International Symposium. Raven Press, New York, pp 141–144

Vayer P, Mandel P, Maitre M (1987) γ-Hydroxybutyrate, a possible neurotransmitter. Life Sci 41: 1547–1557

Vergnes M, Marescaux C, Micheletti G, Depaulis A, Rumbach L, Warter JM (1984) Enhancement of spike and wave discharges by GABA-mimetic drugs in rat with spontaneous petit mal-like epilepsy. Neurosci Lett 44: 91–94

Vergnes M, Marescaux C, Micheletti G, Rumbach L, Warter JM (1985) Blockade of antiabsence activity of sodium valproate by THIP in rats with petit mal-like seizures. J Neural Transm 63: 133–141

Vergnes M, Marescaux C, Depaulis A, Micheletti G, Warter JM (1987) Spontaneous spike and wave discharges in thalamus and cortex in a rat model of genetic petit mal like seizures. Exp Neurol 96: 127–136

Vergnes M, Marescaux C, Depaulis A, Micheletti G, Warter JM (1990) The spontaneous spike and wave discharges in wistar rats: a model of genetic generalized non-convulsive epilepsy. In: Avoli M, Gloor P, Kostopoulous G, Naquet R (eds) Neurobiological approaches. Birkhäuser, New York, pp 238–253

Williams D (1953) A study of thalamic and cortical rhtythms in petit mal. Brain 76: 50–69

Zouhar A, Mares P, Brozek G (1980) Electrocorticographic activity elicited by metrazol during ontogenesis in rats. Arch Int Pharmacodyn Ther 248: 280–288

Author's address: O. Carter Snead III, M.D., Division of Neurology, Children's Hospital, Los Angeles, Box #82, 4650 Sunset Boulevard, Los Angeles, California 90027, U.S.A.

J Neural Transm (1992) [Suppl] 35: 21–36

The tottering mouse: a critical review of its usefulness in the study of the neuronal mechanisms underlying epilepsy

G. K. Kostopoulos

Department of Physiology, Medical School, University of Patras, Patras, Greece

Summary. The tottering mouse resulted from a recessively inherited, autosomal, single-locus mutation which produces a very characteristic neurological and cellular phenotype. Almost simultaneously and late in the development of this mutant appears a triad of symptoms: frequent episodes of absence seizures with spike-and-wave discharges; more rarely occurring episodes of focal motor seizures; and ataxia. Electrographic, behavioural and pharmacological similarities to absence epilepsy in man make the tottering mouse a useful animal model for testing new anti-absence drugs. It also affords a unique opportunity to study the effects of multiple alleles on epileptic behaviour. The neuronal mechanisms underlying the generation of absence seizures in this mutant are apparently a combination of a generalized noradrenergic hyperactivity in the brain and some gene-linked, but unknown, conditions prevailing in an earlier phase of development at specific brain areas which induce the generalized forebrain hyper-innervation by locus coeruleus terminals. Several biochemically, microscopically and electrophysiologically identified cellular differences between normal and tottering mice are potential aspects of this primary developmental defect. Research into these gene-linked neuronal characteristics co-inherited with seizures in this mutant makes the tottering mouse a powerful tool in the study of cellular mechanisms underlying genetically determined factors in epileptogenesis.

Introduction

Studies of mutant mice are increasingly contributing to our efforts to understand some of the mechanisms underlying the epilepsies. At present there are 18 identified loci, on 10 different chromosomes of the mouse genome, whose expressed neurological phenotype includes different seizure patterns (Noebels, 1986; Oiao et al., 1989). The most extensively studied of these mutants is the tottering mouse (Noebels, 1979, 1984a; Noebels and Sidman, 1979; Kaplan et al., 1979), which is the result of a single-locus mutation on autosomal chromosome 8 and is inherited recessively. It has also become fascinating to study and holds out the greatest potential help towards an understanding of epilepsy, since recent findings concerning its genetic background and its specific aberrations at the level of neuronal synaptic transmission indicate many ways in which it can serve as a useful experimental tool.

Neurophysiological phenotype

Homozygous (tg/tg) mice are distinguished by the appearance three to four weeks after birth of the characteristic triad: (a) frequent non-convulsing seizures of the absence type; (b) rarer focal motor seizures (Noebels and Sidman, 1979; Kaplan et al., 1979); and (c) mild hind-limb ataxia (recognized first, hence the name tottering) (Green and Sidman, 1962).

The *non-convulsing seizures* consist of spontaneous sudden arrest of movement and a fixed staring posture, often accompanied by twitching of vibrissae or jaw. Photic stimulation does not provoke these episodes, but they can be blocked by different sensory stimuli. In absolute coincidence with this pattern of behaviour, the electrocorticogram shows stereotypic, generalized and bilaterally synchronous 6/sec spike-and-wave (SW) or polyspike discharges (Noebels and Sidman, 1979; Kaplan et al., 1979).

We studied these animals with chronically implanted EEG electrodes (Kostopoulos et al., 1987). Left unattented and unrestrained, the animals displayed SW — absence seizure episodes with a mean duration of 2.5 sec (range 2 to 10 sec). There was no postictal depression. Instead, the return to normal cortical rhythms and behaviour was as sudden as the onset of the episode.

The frequency of appearance of absence seizures is reported to increase from one per hour on postnatal day 18 to an adult rate of 40–60/hour (Noebels and Sidman, 1979). Thus, about 10% of the resting EEG activity of the tottering mouse consists of SW. In our studies, both the rate of appearance and the duration of SW declined slightly, but significantly, during the course of control recording sessions (standard: 9 a.m. to 1 p.m.) to an average of 35 SW/hour (Fig. 1, Kostopoulos et al., 1987).

Since the level of arousal has an apparent influence on the incidence of SW (Kellaway et al., 1990), we decided to see how caffeine, the most commonly used central stimulant, affects absence seizures in the tottering mouse. Relatively low doses of caffeine (5, 10, 15 mg/kg; n = 8) injected intrapertoneally (i.p.) at the fourth hour of an 8-hour recording session significantly (p < 0.001 for 10 and 15 mg) decreased the incidence of absence seizures, as compared with control saline injections (Fig. 1). During the 30-min post-injection period, SW were eliminated and they reached 50% pre-injection levels only after 1–2 hours.

The observation that amphetamine did not block SW in tg/tg mice (Fig. 1B) is difficult to explain. It suggested, however, that caffeine did not act merely as a general CNS stimulant. A unique property of caffeine

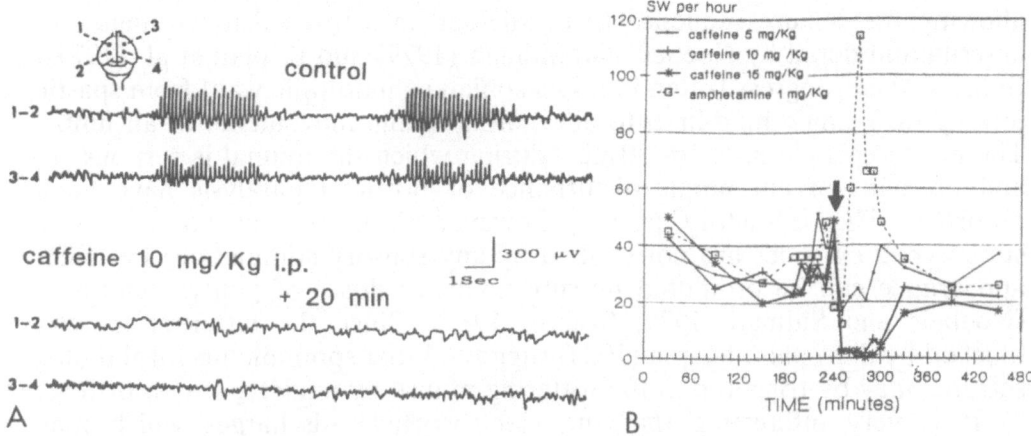

Fig. 1. Intraperitoneal injections of caffeine stop absence seizures with EEG spike and wave discharges (SW) in freely moving tottering mice. Electrocortical bipolar recordings (see inset) from a spontaneously epileptic tg/tg mouse before (upper) and 20 min after (below) the injection of caffeine. Mean incidence of SW in 8 tg/tg mice shows that the effect is significant and dose-dependent ($p < 0.001$ for 10 mg compared to saline control; not shown). This effect of caffeine is not mimicked by an equally arousing dose of amphetamine. Arrow = time of injections (data from Kostopoulos et al., 1987)

among central stimulants is the specific blockade it exerts on adenosine receptors (Phillis and Kostopoulos, 1975; Snyder and Sklar, 1984), and caffeine may somehow act through these receptors.

The above-mentioned observations (Noebels and Sidman, 1979; Kostopoulos et al., 1987) and the lack of SW in tg/tg mice during sleep (Noebels, 1984a), as well as the report (Kaplan et al., 1979) of highly reproducible appearance of SW in tg/tg during an induced state of drowsiness, suggest that absence seizures in tottering mice might have a relation to vigilance and even a circadian rhythmicity similar to that observed in rats, cats and humans (see ref. in Kellaway et al., 1990; Vergnes et al., 1990). Appropriately designed long-term recordings should be conducted in tg/tg mice to answer this interesting question, not only to the benefit of epilepsy research, but also with regard to the effects such an excessive NA cortical innervation (Levitt and Noebels, 1981) might have on sleep-wakefulness patterns.

Absence seizures in adult tg/tg are abolished by ethosuximide, but not by phenytoin (Heller et al., 1983), a fact emphasizing the similarities between this model and absence seizures in humans.

Intermittent *focal motor seizures* also appear spontaneously about once per day, likewise starting 3–4 weeks postnatally and declining in frequency after the sixth month. The seizure pattern is remarkably stereotyped in all animals, including the first appearing episodes in young animals, thus allowing the seizure sequence to be divided into three distinct stages, as described in detail by Noebels and Sidman (1979) and Kaplan et al. (1979). In brief, they progress (often in a Jacksonian somatotopic way) from spastic jerking of a single hind-limb to alternating clonic movements of all limbs. The entire episode lasts 10–30 min, during which the animal is responsive, and no signs of autonomic disturbance or postictal paralysis have been reported. The refractory period, however, lasts several hours. These stereotyped episodes are not evoked by any sensory stimulation, but their appearance can be promoted by subconvulsive doses of pentylenetetrazol (Noebels and Sidman, 1979; Syapin, 1983). Since the latter effect was blocked by diazepam, Syapin (1983) suggested that spontaneous focal motor seizures may by triggered in the tottering mouse by anxiety, stress, or fear.

It is very interesting that the electrocortical discharges which may accompany such stereotype seizure episodes do not necessarily correlate in time with motor behaviour. They may include high-voltage generalized spikes or sharp waves, or periods of desynchronization interspersed with long runs of generalized theta waves; but they do not follow any reproducible pattern, being highly variable or sometimes unremarkable. It may be relevant to note here that 2-D-glucose studies during focal motor seizures in tg/tg mice have revealed higher metabolic rates at certain subcortical diencephalic and brainstem nuclear regions (Noebels and Sidman, 1979). A subcortical primary origin of these seizures may thus be suggested. The anatomical restriction of the increased metabolic activity to certain areas only, without spread to surrounding brain, underlines the focal nature of these seizures (Noebels, 1979).

Although a list of similarities between the tottering focal motor seizures and human simple partial seizures has been offered (Syapin, 1983), more studies are needed before these motor seizures can be categorized in any useful manner; especially as the relatively rare spontaneous occurrence of focal motor seizures in tg/tg mice has not allowed any conclusions with regard to their sensitivity to anti-epileptic drugs (Heller et al., 1983).

The similarities of "absence" seizures with SW in tottering mice to human absence epilepsy and their temporal independence from the motor seizures in the same animal (Kaplan et al., 1979; Kostopoulos et al., 1987) make it unlikely that the tottering SW are interictal manifestations of the motor seizures. Instead, it appears that the tottering mice experience two independent ictal events: the "absence" and the focal motor seizures (Kaplan et al., 1979; Noebels and Sidman, 1979).

Other characteristics of the tottering homozygous mice besides its seizures have not been extensively studied. Along with its above-mentioned wobbly gait, its swimming ability, open-field behaviour and balance on a rotorod were found to be relatively poor (Syapin, 1982). The adult tg/tg weights less than normal and is fertile, although females are such unreliable breeders that we always had to cross male tg/tg with female tg/+ or +/+. A microscopic study of certain hypothalamic areas in this mouse might therefore prove revealing.

Cellular phenotype

The triad of symptoms described above appears relatively late in development, at about the same time as NA axons arising from locus coeruleus (LC) attain full cortical terminal arborization and transmitter synthesis capacity (Elias et al., 1982). This is apparently not coincidental. Levitt and Noebels (1981) discovered that in this mutant there is a gene-linked *proliferation of NA axon terminal arborizations* in certain brain target areas of LC, including hippocampus, cerebellum, thalamus and less so, but significantly, in neocortex. Other catecholaminergic systems appear not affected (Levitt and Noebels, 1981). NA hyperinnervation is already evident from the 5th (30% up) and the 14th (100% up) postnatal day, well before the appearance of SW at 3–4 weeks (Phillips and Levitt, 1986). This proliferation of NA terminals is not accompanied by any change in neuronal number or size (Levitt and Noebels, 1981; Levitt, 1988) or area distribution (Stanfield, 1989b) in nucleus LC, or any increase in number of LC neurons projecting contralaterally or bilaterally (Stanfield, 1989b). Therefore the defect in tg/tg mice concerns rather selectively the development of LC terminal arbors within certain brain regions (Levitt, 1988). Neither does it reflect a shrinkage of these target regions (Noebels and Sidman, 1979): if NA innervation is abolished shortly after birth, absence episodes with SW do not develop in the adult tottering mouse (Noebels, 1984a).

NA levels (Levitt and Noebels, 1981) and NA metabolite 3-methoxy 4-hydroxyphenylglycol (Heller, 1984) are found increased by 30–200% in the cortical LC target areas of the adult tg/tg mouse, while in pons NA levels are decreased (Abbott et al., 1988). Since NA receptor (α and β) numbers and affinity are not down-regulated in tg/tg cortex (Dusser and Peroutka, 1990), hippocampus and cerebellum (Levitt et al., 1987), and the turnover of NA per NA terminal axon is found normal (Levitt, 1988), we are apparently dealing with a genuine gene-linked increase in NA influence on target neurons. At the metabolic level however, a decreased glycogenolytic potency of NA (50 times higher threshold concentration, postsynaptic perhaps β-NA receptor effect) in tg/tg neocortex has already been documented (Magistretti et al., 1987) from the 1st postnatal day (Magistretti and Hof, 1987). The study unfortunately does not address the question of how specific this subsensitivity is for NA. In a different study, cortical slices from adult tg/tg mice were found to have elevated basal cAMP levels compared to those of control mice (Jalilian Tehrani and Barnes, 1990). This difference was abolished by isoprenaline, while the ability of the slices to produce cAMP upon NA β-receptor stimulation was not different in the two groups. It was therefore concluded that a higher level of endogenous NA release in the tg/tg mice contributes to an elevation of basal cAMP levels.

We know relatively less about other transmitter systems besides catecholamines in tottering mice. Choline acetyl transferase activity is not abnormal (Psarropoulou et al., 1987). The density — but not the affinity —

of muscarinic receptors is relatively higher on the 10th postnatal day in tg/tg hippocampus and cortex, but later (22nd day) it is already significantly lower, by 40–60% (Liles et al., 1986). A recent report, however, shows a normal adult tg/tg brain binding to muscarinic receptors as well as to receptors for 5-hydroxytryptamine and dopamine (Dusser and Peroutka, 1990). $GABA_A$ and $GABA_B$ receptors in tg/tg cerebellum are normal as well as benzodiazepines binding (Psarropoulou et al., 1987). Finally, relatively lower levels of reduced and oxidized glutathione were found in cortical structures of the tg/tg brain (Abbott et al., 1990). This observation may prove of significance, as reduced glutathione is the most important compound that can prevent membrane lipid peroxidation, and the latter has been causally implicated in some forms of epilepsy (Abbott et al., 1990).

Thorough histopathological examination by Noebels and Sidman (1979) and Noebels (1986) did not confirm earlier claims of minor cellular lesions in cerebellum (Meier and McPike, 1971; Syapin, 1982; Seyfried et al., 1981) supposedly underlying the ataxia in tg/tg mice. The significance of those earlier observations is also diminished by the fact that they were made only in adult mice after the phenotype had developed. A subtle anatomical difference of possibly direct relevance to epilepsy was discovered in tg/tg hippocampus (Stanfield, 1989a): there is an excessive expression of mossy fibers in the infra- and supragranular layers of the temporal dentate gyrus in tg/tg compared to tg/+ mice.

Recent in vitro electrophysiological studies (see Kostopoulos and Psarropoulou, present volume) have demonstrated that hippocampal neurons of tottering mice have abnormally high postsynaptic excitability and altered sensitivity to extracellular potassium as compared to their normal litter-mates (Kostopoulos et al., 1988; Kostopoulos and Psarropoulou, 1990).

On the epileptic nature of the tottering mouse spike-and-wave discharges

It has been debated whether the 6/sec SW discharges in tottering mice represent epileptiform discharges or enhanced spindle activity (Kaplan, 1985). It is crucial for the validity of using the tottering mutant as a model of generalized epilepsy in man to note the following with regard to this matter.

(a) A resemblance of SW to large EEG spindles with regard to their relationship to vigilance and with regard to the respective intraburst frequencies is also found in humans with absence seizures (Kellaway et al., 1990) and in animal models of this disorder (Gloor et al., 1979, 1990; Kostopoulos et al., 1981; Avoli et al., 1983; Kostopoulos and Gloor, 1982). In these reports, the resemblances between spindles and SW are actually part of the argument supporting a proposed way of development of SW epilepsy through potentiation of neuronal mechanisms underlying spindles.

(b) It is true that some kinds of rhythmical EEG activity associated with drowsiness or arrest of movements may be part of the behavioural repertory of normal rodents and should not necessarily be equated to epilepsy (Kaplan, 1985; Buzsaki et al., 1988). However, the episodes of SW in tottering mice are highly uniform, appear much more regularly at rates exceeding 1000/day and have greater bilateral synchrony than spindling (Noebels and Sidman, 1979; Kostopoulos et al., 1987), which in rats has higher frequency, rather 7–10 (Hammond et al., 1979).

(c) While spindling activity occurs with increased probability after the animal has assumed a sleep position, SW in tottering mice have been shown in combined EEG and videotape recordings to start simultaneously — interrupting self-paced motor behaviour — and absolutely coincide in time with a behaviour resembling human absence (Heller et al., 1983). Return to normal EEG and behavioural activity is as sudden as its initiation.

(d) The appearance of typical SW-absence episodes of the tottering mouse is gene-linked, being restricted only in the homozygous mice (tg/tg) and never appearing in +/+ or tg/+ animals (Noebels and Sidman, 1979). As explained below, absence seizures in tg/tg are co-inherited with another type of epilepsy, this one of the focal motor type (Noebels and Sidman, 1979). Furthermore, in certain alleles of the same locus as the tg/tg^{la}, both these types of experimental epilepsy are intensified (Noebels, 1984b; Seyfried and Glaser, 1985).

(e) An observation clearly distinguishing SW from spindling activity in tottering mice is that SW appeared to be able to interrupt and override the motor seizures of the mouse and even to have an exceptionally higher rate of appearance at a specific — generalized — stage of these stereotypic motor seizures (Kostopoulos et al., 1987).

(f) There is evidence suggesting that SW are associated to a gene-linked hyperexcitability in tottering mice. During some SW episodes, single sudden myoclonic jerks of a limb or the trunk, or a sudden drop of the head corresponded with EEG spikes of larger amplitude (Noebels and Sidman, 1979). Myoclonic jerks have been taken as an indication of hyperexcitability (Kaplan, 1985). Neuronal hyperexcitability has been demonstrated in vitro in hippocampal slices from tottering mice (Kostopoulos and Psarropoulou, 1990). Finally the excessive expression of mossy fibers and the decrease in muscarinic receptors, both appearing in tottering hippocampus, have been reported in other animal models to result from convulsive or otherwise induced neuronal hyperexcitability (Stanfield, 1989a; Liles et al., 1986).

(g) Ethosuximide is very effective in blocking SW in tottering mice, and generally the profile of responsiveness of the SW discharge to anticonvulsant drugs in tottering mice is similar to that found in human absence epilepsy (Heller et al., 1983).

(h) Besides the neocortex, SW are also recorded in tottering hippocampus (Noebels, 1984a), while spindles are primarily a thalamocortical rhythm.

(i) Finally, neonatal 6-OHDA lesions of locus coeruleus cause a complete and specific disappearance of SW-absence episodes in the adult tg/tg mutant, without otherwise altering the background EEG rhythms of freely behaving mice (Noebels, 1984a).

The above arguments could suffice to justify the use of the tottering mouse as a model of generalized absence seizures. As pointed out by Kaplan (1985), however, definite proof of the epileptic nature of the recorded SW episodes is still lacking; it would require evidence that the animals during the SW were not conscious of changes in their environment, as demonstrated in other animal models (Taylor-Courval and Gloor, 1984). What has been observed is that SW episodes in tottering mice can be interrupted by sensory stimulation (Noebels and Sidman, 1979). An appropriate response to the stimulus, however, is not elicited. Well-designed psychological experiments are certainly needed to clarify this point.

Putative mechanisms of epileptogenesis in the tottering mouse

The pivotal finding in the tg/tg research has been the demonstration that in this mutant absence seizures are co-inherited and develop at the same time as a profound NA hyperinnervation of certain parts of the mouse brain (Levitt and Noebels, 1981). However, the way in which the two parts of the phenotype are related to each other remains obscure. It appears fundamental to answer the following questions. (a) Is NA hyperinnervation the direct cause of absence seizures? (b) Is NA hyperinnervation the primary gene-linked defect, or is it a secondary adaptation to an NA subsensitivity? In the latter case, what is the primary defect, and how does it contribute (if it does so at all) to the development of seizures. The present state of knowledge, especially of the cellular phenotype of the tottering mouse, allows little more than a refinement of these questions, so that they will be addressed experimentally.

(a) Regarding the *role of NA hyperinnervation in tottering absence seizures*, it should be noted that in most other animal models of epilepsy the evidence favours an anticonvulsant role of NA (Mason and Corcoran, 1979). Drugs which increase NA transmission decrease SW in a rat genetic model of generalized epilepsy (Micheletti et al., 1987). Caffeine, which can enhance NA release (Jonzon and Fredholm, 1984) and activate LC neurons (Olpe, 1982; Shefner and Chiu, 1986), blocks SW in tottering mice (Kostopoulos et al., 1987). There are other mutants, controlled by other chromosomes than those in tg/tg mice, which display absence seizures with SW (Oiao et al., 1989; Noebels, 1986). Not all of them show NA hyperinnervation (star-gazer mutant, Oiao et al., 1989), while several mutants with NA hyperinnervation do not suffer from absence seizures (Noebels, 1986; Seyfried and Glaser, 1985). Finally, while the absence seizures in tg/tg start at the developmental time of peak maturation of cortical NA transmission, in the compound heterozygote allele tg/tg[la] they start two weeks earlier (see Noebels, 1984b). This suggests the existence of another gene-linked defect, also contributing to the mechanisms underlying absence seizures, and which matures earlier than NA transmission. It becomes, therefore, clear that the relationship between NA hyperinnervation and absence seizures is not likely to be simply causal and exclusive.

However, tottering absence seizures do not develop if the arborization of NA axons is prevented by neonatal administration of 6-OH-DA (Noebels, 1984a). Furthermore, unilateral destruction of LC in adult tg/tg mice reduces the rate of appearance of SW and disrupts their bilateral synchrony. This is solid evidence that NA makes a positive contribution to the mechanisms underlying absence seizures in this mouse. Also, the alpha-2 adrenoreceptor agonist clonidine prolongs the duration of SW and absence episodes in tg/tg (from 2.6 to 13 sec, Heller, 1984). An obvious problem with this view is that NA is expected to have a desynchronizing effect on EEG (McCormick and Prince, 1988, see below), while SW demand a great deal of electrocortical synchronization. Even though the NA effect is necessary for the development of SW, it could not produce it by itself — and it does not (Micheletti et al., 1987). Taking into account all the reviewed experiments in tottering mice, one can at least conclude that the neuronal mechanisms underlying absence seizures in tottering mice should be searched for among those mechanisms which are drastically — even though not exclusively — modulated by excessively produced NA.

(b) The *question of primacy between pre- and postsynaptic mechanisms* would seem to be most appropriately dealt with in developmental studies demonstrating the exact sequence of appearence of each of the known cellular aberrations in relationship to the development of seizures in tg/tg mice. As reported above, the relative insensitivity to the glycogenolytic effect of NA apppears from the first day of life of the tg/tg mouse, proliferation of LC terminals starts from the 5th postnatal day and absence seizures start three weeks later. In view of this sequence of events, and since a decrease in both glycogen levels (in DBA/2J mice, Schreiber, 1981) and NA levels (PTZ seizures, Burley and Ferrendelli, 1984) increases susceptibility to certain epilepsies, Megistretti et al. (1987) suggested that in tg/tg a decrease in the efficacy and potency of NA to mobilize energy substrates may generate conditions that favour the expression of seizure activity, while causing a secondary proliferation of NA terminals. It should be noted that in all cases where NA hyperinnervation was the primary (experimentally induced) effect, a down-regulation of NA receptors followed (Rosenblatt et al., 1979; Harden et al., 1979; Beaulieu and Coyle, 1982). Interestingly, this is not the case in tg/tg (Levitt et al., 1987). Especially supportive of a secondary adaptive nature of NA hyperinnervation in tg/tg mice is its marked area specificity. Local conditions seem to selectively affect the development of LC terminal arbors within certain brain regions and not in others (Stanfield, 1989b).

Since NA hyperinnervation starts well before the maturation of synaptic transmission, the signal for this adaptation is not likely to be the deficiency of its synaptic receptors. It could be a disruption during development of the interaction between axonal growth and the retrograde transport of growth factors (see Stanfield, 1989b). The reduced glycogenolytic response to NA may be part of such a scheme, along with the numerous other cellular defects (see "cellular phenotype" above), if they also prove to take place so early in development. Indirect evidence that these postsynaptic defects may by themselves promote epilepsy has been reported (Magistretti et al., 1987; Abbott et al., 1990; Stanfield, 1989a).

One may therefore conclude that NA hyperinnervation in tottering mice is contributory to their absence seizures, while being secondary to a primary postsynaptic defect occurring earlier in development. The combination of the two defects finally leads to SW. The question is, how.

(c) Any *attempted synthesis* of the above has to cross-examine the various effects of NA and the known and suspected neuronal mechanisms underlying the development of SW. NA has been demonstrated to affect neuronal excitability in several possible ways, depending on the diverse properties and uneven brain distribution of its different receptors (see reviews by Aghajanian, 1984; Nicoll et al., 1990). Most prominent among these effects are: a slow depolarization due to decrease in K^+ conductance through α_1-receptors, a slow hyperpolarization (I_{AHP}) due to increase in K^+ conductance through α_2-receptors, and, finally, a blockade of slow after-hyperpolarization through β-receptors. In general, there is agreement between electrophysiological and autoradiographic studies that α-receptors are most often found in subcortical areas and β-receptors in cortex, with the — important for our subject — predominance of α_1-receptors in thalamus and α_2-receptors on LC neurons.

The study of several experimental models has suggested that SW may develop when rhythmical thalamocortical volleys responsible for the pacing of spindles are received by hyperexcitable cortical neurons (Steriade, 1990; Gloor et al., 1990; Kostopoulos and Gloor, 1982). Both conditions seem to be necessary, while NA is able to affect SW development at both cortical and thalamic ends, albeit in opposite ways. At the thalamic site NA can effectively inhibit the generation of thalamocortical rhythms primarily by an α_1-receptor-mediated inactivation of low threshold Ca^{2+} spikes (LTC, McCormick and Prince, 1988). Note that anti-absence drugs like etho-suximide — which abolish SW also in tg/tg mice — reduce LTC (Coulter et al., 1990). At the cortical level, NA can enhance excitability and poten-tiate the cortical response to thalamic input by reducing spike frequency adaptation through its β-receptors (Foehring et al., 1989). Any change in the balance which would favour the latter effect of NA, could conceivably promote SW. A tottering mutation which through an unknown postsynaptic mechanism renders NA α_1-receptors ineffective in thalamus — thus allowing thalamocortical synchronization — while at the same time triggering a reactive NA hyperinnervation in the entire forebrain — thus enhancing cortical excitability — would promote SW. It would also be consistent with both the secondary appearance of generalized NA hyperinnervation and the requirement of LC terminals for the development of SW in this animal.

Putative gene-linked postsynaptic defects could include an altered neuronal excitability and/or responsiveness to high extracellular K^+, both demonstrated for tottering hippocampal neurons in vitro (see Kostopoulos and Psarropoulou, present volume). It is interesting to note that NA levels in tg/tg brain were found to be much more increased in dorsal thalamus (175%) than in cortex (27%, Levitt and Noebels, 1981), and that a selective and marked increase in metabolic activity of several thalamic nuclei is revealed during tg/tg motor seizures (Noebels and Sidman, 1979). The reported lower NA levels in tg/tg brain stem (Abbott et al., 1988) may reflect less effective negative feedback through α_2-NA receptors on LC neurons, which may further enhance SW.

It is obvious that the above constitutes only one of the many hypotheses which the present state of knowledge allows us to make in order to explain the mechanisms underlying the development of absence seizures in the tottering mouse. It tries to see the tottering mouse under the light of recent developments in research into generalized epilepsy (Avoli et al., 1990), which emphasize the role of thalamic LTC and cortical hyperexcitability in the genesis of SW. Although speculative, the hypothesis is testable. It would be consistent with a LTC relatively more prominent and/or less affected by NA in tottering thalamocortical neurons. An in vitro study of tottering thalamic neurons might therefore be revealing, particularly as some electrophysiological abnormalities have been already shown to be co-inherited with epilepsy in this animal and to be maintained in vitro (Kostopoulos and Psarropoulou, present volume). Histopathological studies should focus on tg/tg thalamus and especially on its thalamocortical and reticular neurons. Finally, the conspicuously lacking electrophysiological experiment is the recording from identified LC neurons during tottering absence seizures.

Usefulness of the tottering mouse as a model of epilepsy

Theoretically, the tottering mutant could be classified either under models of absence seizures or under models of partial seizures (Fisher, 1989), since its phenotype includes characteristics of two independently occurring, but co-inherited types of seizure. Clinical absence is characterized by arrest of ongoing activity, partial or full decrease of awareness and concurrent 3–4/sec SW in the EEG (Dreifuss, 1990). With the limitations imposed to greater or lesser extent on all animal models of a human disease, the *absence seizures* of tg/tg mice resemble the corresponding human condition electrographically (Noebels and Sidman, 1979; Kaplan et al., 1979; Kostopoulos et al., 1987), behaviourally (Heller et al., 1983; Noebels and Sidman, 1979) and pharmacologically (Heller et al., 1983) and they display a similar sensitivity to arousal and caffeine (Kostopoulos et al., 1987). The steady spontaneous rate at which SW-absence episodes appear in the homozygotes only of these mice offers a very convenient model for testing new anti-absence drugs.

The co-inherited *partial motor seizure* pattern provides a less useful epileptic model because (a) its appearance is very infrequent and cannot be reliably provoked by any physical stimulus, and (b) it has not been possible to characterize these episodes electrographically (Noebels, 1986). However, the dramatically more severe expression of seizures (both absence and focal motor) in the two known alleles of tg/tg, leaner and roller, affords the opportunity for assessing the biochemical and physiological action of multiple alleles on epileptic behaviour. In particular, the focal motor seizures in tg/tg[la] offer, to our knowledge, the only natural animal model of status epilepticus (Seyfried and Glaser, 1985).

Neither the cellular phenotype of tg/tg mice, nor the simple mendelian way of their inheritance characterizes human absence seizures. However, *it is precisely at these two levels of analysis — the cellular and the genetic — that the tottering mouse holds out the greatest potential* to help us understand epileptic mechanisms. A single gene mutation with a single (possibly modifiable) molecular disturbance at its core can generate a complex, but rather stereotypical functional disturbance involving extensive parts of the nervous system and presents a phenotype which resembles human petit mal. An example of the powerful genetic-engineering tools now available in this respect is the recent discovery of a tottering gene-linked ectopic expression of a NA-synthesizing enzyme (Hess and Wilson, 1989, see also Zhu et al., 1989). Also, the electrophysiological arsenal recently developed for the in vitro hippocampal slice (Dingledine, 1985) can be exploited to the benefit of epilepsy research, once a gene-linked postsynaptic hyperexcitability and other electrophysiological characteristics related to epileptogenesis are found to be maintained in this convenient in vitro preparation (see Kostopoulos and Psarropoulou, 1990 and present volume).

Understanding the mechanisms which underlie the development of epileptic synchronization in the form of SW goes beyond its appearance in absence seizures, since there is evidence that SW may actually be a marker of genetic predisposition to perhaps all types of epilepsy (Gloor, 1979, 1982).

Acknowledgement

The studies described were supported by grants from the Greek Ministry of Research and Technology.

References

Abbott LC, Weber B, Abhold RH (1988) Norepinephrine and met-encephalin concentrations in specific brainstem and spinal cord regions of the genetically epileptic tottering (tg/tg) mouse. Soc Neurosci Abstr 14: 830

Abbott LC, Nejad HH, Bottje WG, Hassan AS (1990) Glutathione levels in specific brain regions of genetically epileptic (tg/tg) mice. Brain Res Bull 25: 629–631

Aghajanian GK (1984) The physiology of central alpha- and beta-adrenoreceptors. In: Usdin E, Carlson A, Dahlstrom A, Engel J (ed) Catecholamines: neuropharmacology and central nervous system. Theoretical aspects. Liss, New York, pp 85–92

Avoli M, Gloor P, Kostopoulos G, Gotman J (1983) An analysis of penicillin-induced spike and wave discharges using simultaneous recordings of cortical and thalamic single neurons. J Neurophysiol 50: 819–837

Avoli M, Gloor P, Kostopoulos G, Naquet R (eds) (1990) Generalized epilepsy: neurobiological approaches. Birkhäuser, Boston, pp 480

Beaulieu M, Coyle JT (1982) Fetally-induced noradrenergic hyperinnervation of cerebral cortex results in persistent down-regulation of beta-receptors. Dev Brain Res 4: 491–494

Burley ES, Ferrendelli JA (1984) Regulatory effects of neurotransmitters on electroshock and pentylenetetrazol seizures. Fed Proc 43: 2521–2524

Buzsaki G, Bickford RG, Ponomareff G, Thal LJ, Mandel R, Gage FH (1988) Nucleus basalis and thalamic control of neocortical activity in the freely moving rat. Neuroscience 8(11): 4007–4026

Coulter DA, Huguenard JR, Prince DA (1990) Cellular actions of petit mal anticonvulsants: implication of thalamic low-threshold calcium current in generation of spike and wave discharges. In: Avoli M, Gloor P, Kostopoulos G, Naquet R (eds) Generalized epilepsy: neurobiological approaches. Birkhäuser, Boston, pp 425–435

Dingledine R (1984) Brain slices. Plenum Press, New York

Dreifuss FE (1990) The syndromes of generalized epilepsy. In: Avoli M, Gloor P, Kostopoulos G, Naquet R (eds) Generalized epilepsy: neurobiological approaches. Birkhäuser, Boston, p 480

Dusser AE, Peroutka SJ (1990) Neurotransmitter receptors in adult tottering (tg/tg) mice. Epilepsia 31(4): 378–381

Elias M, Deacon T, Caviness VS Jr (1982) The development of neocortical adrenergic innervation in the mouse: a quantitative radioenzymatic analysis. Dev Brain Res 3: 652–656

Fisher RS (1989) Animal models of the epilepsies. Brain Res Rev 14: 245–278

Foehring RC, Schwindt PC, Crill WE (1989) Norepinephrine selectively reduces slow Ca^{2+} and Na^+-mediated K^+ currents in cat neocortical neurons. J Neurophysiol 61: 245–256

Gloor P (1979) Generalized epilepsy with spike-and-wave discharge: a reinterpretation of its electrographic and clinical manifestations. Epilepsia 20: 571–588

Gloor P (1982) Toward a unified concept of epileptogenesis. In: Akimoto H, Kazamatsuri H, Seino M, Ward A (eds) Advances in Epileptology: XIIIth Epilepsy International Symposium. Raven Press, New York, pp 83–86

Gloor P, Pellegrini A, Kostopoulos GK (1979) Effects of changes in cortical excitability upon the epileptic bursts in generalized penicillin epilepsy of the cat. Electroencephalogr Clin Neurophysiol 46: 274–289

Gloor P, Avoli M, Kostopoulos G (1990) Thalamocortical relationships in generalized epilepsy with bilaterally synchronous spike-and-wave discharge. In: Avoli M, Gloor P, Kostopoulos G, Naquet R (eds) Generalized epilepsy: neurobiological approaches. Birkhäuser, Boston, pp 190–212

Green MC, Sidman RL (1962) Tottering: a neuromuscular mutation in the mouse. J Hered 53: 233–237

Hammond EJ, Villarreal HJ, Wilder BJ (1979) Distinction between normal and epileptic rhythms in rodent sensorimotor cortex. Epilepsia 20: 511–518

Harden TK, Mailman RB, Mueller RA, Breese GR (1979) Noradrenergic
    hyperinnervation reduces the density of b-adrenergic receptors in rat cerebellum.
    Brain Res 166: 194–198
Heller AH (1984) Clonidine exacerbates absence seizures in the mutant mouse
    tottering. Soc Neurosci Abstr 10: 411
Heller AH, Dichter MA, Sidman RL (1983) Anticonvulsant sensitivity of absence
    seizures in the tottering mutant mouse. Epilepsia 25: 25–34
Hess EJ, Wilson MC (1989) Tyrosine hydroxylase is expressed in the purkinje cells of
    the allelic mouse mutants tottering and leaner. Soc Neurosci Abstr 15: 986
Jalilian Tehrani MH, Barnes EM Jr (1990) Basal and drug-induced cAMP levels in
    cortical slices from the tottering mouse. Epilepsy Res 7: 205–209
Jonzon B, Fredholm BB (1984) Adenosine receptor mediated inhibition of
    noradrenaline release from slices of the rat hippocampus. Life Sci 35: 1971–1979
Kaplan BJ (1985) The epileptic nature of rodent electrocortical polyspiking is still
    unproven. Exp Neurol 88: 425–436
Kaplan BJ, Seyfred TN, Glaser GH (1979) Spontaneous polyspike discharges in an
    epileptic mutant mouse (tottering). Exp Neurol 66: 577–586
Kellaway P, Frost JD, Crawley JW (1990) The relationship between sleep spindles and
    spike-and-wave bursts in human epilepsy. In: Avoli M, Gloor P, Kostopoulos G,
    Naquet R (eds) Generalized epilepsy: neurobiological approaches. Birkhäuser,
    Boston, pp 36–48
Kostopoulos G, Gloor P (1982) A mechanism for spike-wave discharge in feline
    penicillin epilepsy and its relationship to spindle generation. In: Sterman MB,
    Shouse MN, Passouant P (eds) Sleep and epilepsy. Academic Press, New York, pp
    11–22
Kostopoulos G, Psarropoulou C (1990) Increased postsynaptic excitability in
    hippocampal slices from the tottering epileptic mutant mouse. Epilepsy Res 6:
    49–55
Kostopoulos G, Psarropoulou C (1992) Possible mechanisms underlying
    hyperexcitability in the epileptic mutant mouse tottering (this volume)
Kostopoulos G, Gloor P, Pellegrini A, Gotman J (1981) A study of the transition from
    spindles to spike and wave discharge in feline generalized penicillin epilepsy:
    microphysiological features. Exp Neurol 73: 55–77
Kostopoulos G, Veronikis DK, Efthimiou I (1987) Caffeine blocks absence seizures in
    the tottering mutant mouse. Epilepsia 28(4): 415–420
Kostopoulos G, Psarropoulou C, Haas H (1988) Membrane properties, response to
    amines and to tetanic stimulation of hippocampal neurons in the genetically
    epileptic mutant mouse tottering. Exp Brain Res 72: 45–50
Levitt P (1988) Normal pharmacological and morphometric parameters in the
    noradrenergic hyperinnervated mutant mouse tottering. Cell Tissue Res 252: 175–
    180
Levitt P, Noebels JL (1981) Mutant mouse tottering: selective increase of locus
    coeruleus axons in a defined single-locus mutation. Proc Natl Acad Sci USA 78:
    4630–4634
Levitt P, Lau C, Pylypiw A, Ross LL (1987) Central adrenergic receptor changes in the
    inherited noradrenergic hyperinnervated mutant mouse tottering. Brain Res 418:
    174–177
Liles WC, Taylor S, Finnel R, Lai H, Nathanson NM (1986) Decreased muscarinic
    acetylcholine receptor number in the central nervous system of the tottering (tg/tg)
    mouse. J Neurochem 46: 977–982
Magistretti PJ, Hof PR (1987) ^{3}H-Glycogen hydrolysis in the cerebral cortex of two
    spontaneously epileptic mouse mutants: noradrenergic subsensitivity in the
    tottering mouse and age-dependent supersensitive response to K^+ in the quaking
    mouse. Soc Neurosci Abstr 13: 1077

Magistretti PJ, Hof PR, Celio MR (1987) Noradrenergic sub-sensitivity in the cerebral cortex of the tottering mouse, a spontaneously epileptic mutant. Brain Res 403: 181–185

Mason ST, Corcoran ME (1979) Catecholamines and convulsions. Brain Res 170: 497–507

McCormick DA, Prince DA (1988) Noradrenergic modulation of firing pattern in guinea pig and cat thalamic neurons, in vitro. J Neurophysiol 59(3): 978–996

Meier H, MacPike D (1971) Three syndromes produced by two mutant genes in the mouse. Clinical pathological and ultrastructural bases of tottering, leaner and heterozygous mice. J Heredity 62: 297–302

Micheletti G, Walter GM, Marescaux C, Depaulis A, Tranchant C, Rumbach L, Vergnes M (1987) Effects of drugs affecting noradrenergic neurotransmission in rats with spontaneous petit-mal like seizures. Eur J Pharmacol 135: 397–402

Nicoll RA, Malenka RC, Kauer JA (1990) Functional comparison of neurotransmitter receptor subtypes in mammalian central nervous system. Physiol Rev 70(2): 513–565

Noebels JL (1979) Analysis of inherited epilepsy using single locus mutations in mice. Fed Proc 38: 2405–2410

Noebels JL (1984a) A single gene error of noradrenergic axon growth synchronizes central neurons. Nature 310: 409–411

Noebels JL (1984b) Isolating single genes of the inherited epilepsies. Ann Neurol 16 [Suppl]: s18–s21

Noebels JL (1986) Mutational analysis of inherited epilepsies. In: Delgado-Escueta AV, Ward Jr AA, Woodbury DM, Porter RJ (eds) Advances in epilepsy, vol 44. Raven Press, New York, pp 97–113

Noebels JL, Sidman RL (1979) Inherited epilepsy: spike-wave and focal motor seizures in the mutant mouse tottering. Science 204: 1334–1336

Oiao X, Noebels JL, Bronson RT, Davisson MT (1989) Stargazer: a neurological mutant with a complex pattern of inherited spike-wave seizures. Soc Neurosci Abstr 15: 48

Olpe H-R (1982) The locus coeruleus as a target for the activating action of vincamine, nicotine and caffeine. Experientia 38: 757

Phillips E, Levitt P (1986) Developmental expression of the hypertrophied locus coeruleus terminal arbor in the mutant mouse tottering. Soc Neurosci Abstr 12: 1361

Phillis JW, Kostopoulos GK (1975) Adenosine as a putative transmitter in the cerebral cortex. Studies with potentiators and antagonists. Life Sci 17: 1085–1094

Psarropoulou C, Angelatou F, Matsokis N, Veronikis DK, Kostopoulos G (1987) Absence of modification in GABA and benzodiazepine binding and in choline acetyltransferase activity in brain areas of the epileptic mutant mouse tottering. Gen Pharmacol 18(6): 593–597

Rosenblatt JE, Pert CB, Tallman JF, Pert A, Bunnery WE (1979) The effect of imipramine and lithium on a and b receptor binding in the rat brain. Brain Res 160: 186–191

Schreiber RA (1981) Developmental changes in brain glucose, glycogen, phosphocreatine and ATP levels in DBA/2J and C57BL/6J mice and audiogenic seizures. J Neurochem 37: 655–661

Seyfried TN, Glaser GH (1985) A review of mouse mutants as genetic models of epilepsy. Epilepsia 26(2): 143–150

Seyfried TN, Itoh T, Glaser GH, Miyazawa, Yu RK (1981) Cerebellar gangliosides and phospholipids in mutant mice with ataxia and epilepsy: the tottering/leaner syndrome. Brain Res 216: 429–436

Shefner SA, Chiu TH (1986) Adenosine inhibits locus coeruleus neurons: an intracellular study in a rat brain slice preparation. Brain Res 366: 364–368

Snyder S, Sklar P (1984) Behavioral and molecular actions of caffeine: focus on adenosine. J Psychiatr Res 18: 91–106

Stanfield BB (1989a) Excessive intra- and supragranular mossy fibers in the dentate gyrus of tottering (tg/tg) mice. Brain Res 480: 294–299

Stanfield BB (1989b) The distribution of hippocampal and spinal projecting cells in the locus coeruleus of tottering mice. Neuroscience 32: 381–386

Steriade M (1990) Spindling, incremental thalamocortical responses and spike-wave epilepsy. In: Avoli M, Gloor P, Kostopoulos G, Naquet R (eds) Generalized epilepsy: neurobiological approaches. Birkhäuser, Boston, pp 161–180

Syapin PJ (1982) Effects of the tottering mutation in the mouse: multiple neurological changes. Exp Neurol 76: 566–573

Syapin PJ (1983) Inhibition of pentylenetetrazol induced genetically-determined stereotypic convulsions in tottering mutant mice by diazepam. Pharmacol Biochem Behav 18: 389–394

Taylor-Courval D, Gloor P (1984) Behavioural alterations associated with generalized spike and wave discharges in the EEG of the cat. Exp Neurol 83: 167–186

Vergnes M, Marescaux C, Despaulis A, Micheletti G, Warter JM (1990) Spontaneous spike-and-wave discharges in wistar rats: a model of genetic generalized nonconvulsive epilepsy. In: Avoli M, Gloor P, Kostopoulos G, Naquet R (eds) Generalized epilepsy: neurobiological approaches. Birkhäuser, Boston, pp 238–253

Zhu Z, Armstrong DL, Grossman RG, Hamilton WJ (1989) Tyrosine-hydroxylase-immunoreactive neurons in the temporal lobe in complex partial seizures. Ann Neurol 27: 565–572

Author's address: Dr. G. Kostopoulos, Department of Physiology, University of Patras Medical School, Patras, Greece 261 10.

J Neural Transm (1992) [Suppl] 35: 37–69

Genetic absence epilepsy in rats from Strasbourg — A review

C. Marescaux[1], M. Vergnes[2], and A. Depaulis[2]

[1] Service de Neurologie I, C.H.U., and [2] Laboratoire de Neurophysiologie et Biologie des Comportements, Centre de Neurochimie, C.N.R.S., Strasbourg, France

Summary. We have selected a strain of rats and designated it the Genetic Absence Epilepsy Rat from Strasbourg (GAERS). In this strain, 100% of the animals present recurrent generalized non-convulsive seizures characterized by bilateral and synchronous spike-and-wave discharges accompanied with behavioural arrest, staring and sometimes twitching of the vibrissae. Spontaneous SWD (7–11 cps, 300–1,000 µV, 0.5–75 sec) start and end abruptly on a normal background EEG. They usually occur at a mean frequency of 1.5 per min when the animals are in a state of quiet wakefulness. Drugs effective against absence seizures in humans (ethosuccimide, trimethadione, valproate, benzodiazepines) suppress the SWD dose-dependently, whereas drugs specific for convulsive or focal seizures (carbamazepine, phenytoin) are ineffective. SWD are increased by epileptogenic drugs inducing petit mal-like seizures, such as pentylenetetrazol, gamma-hydroxybutyrate, THIP and penicillin.

Depth EEG recordings and lesion experiments show that SWD in GAERs depend on cortical and thalamic structures with a possible rhythmic triggering by the lateral thalamus. Most neurotransmitters are involved in the control of SWD (dopamine, noradrenaline, NMDA, acetylcholine), but GABA and gamma-hydroxybutyrate (GHB) seem to play a critical role. SWD are genetically determined with an autosomal dominant inheritance. The variable expression of SWD in offsprings from GAERS × control reciprocal crosses may be due to the existence of multiple genes.

Neurophysiological, behavioural, pharmacological and genetic studies demonstrate that spontaneous SWD in GAERS fulfil all the requirements for an experimental model of absence epilepsy. As the mechanisms underlying absence epilepsy in humans are still unknown, the analysis of the genetic thalamocortical dysfunction in GAERS may be fruitful in investigations of the pathogenesis of generalized non-convulsive seizures.

Introduction

While searching in the 1980's for a model of partial epilepsy in rats, we recorded electroencephalograms (EEGs) in control rats without any lesion.

We found then that about 30% of the Wistar rats from our breeding colony presented spontaneous spike-and-wave discharges (SWD) which were bilateral and synchronous all over the cortex. During the SWD, the rat is immobile, except for occasional twitches of the face. These episodes appeared similar to the generalized non-convulsive seizures of absence or petit mal epilepsy in humans (Vergnes et al., 1982; Marescaux et al., 1984a,d). In order to demonstrate the validity of this hypothesis, we have undertaken a systematic analysis of this phenomenon.

As only a small proportion of the animals were affected, we tried to increase their numbers by selecting breeders with SWD. A strain in which all animals had SWD in the EEG was obtained in a few generations. This strain was named the "Genetic Absence Epilepsy Rat from Strasbourg" (GAERS) and has now been inbred through 20 generations. Similarly, a control strain free of any spontaneous SWD was outbred over 15 generations.

We report here the results obtained over the past years in neurophysiological, behavioural, pharmacological and genetic studies. These data demonstrate that SWD in the GAERS strain fulfil all the requirements for an experimental model of petit mal seizures (Mirsky et al., 1986; Fariello and Golden, 1987; Snead, 1988). These requirements are: (1) EEG and behavioural similarities with human petit mal seizures, i.e. SWD of 5 to 15 seconds duration on a normal background EEG; (2) arrest of movement and reduced responsiveness; (3) increased occurrence of SWD by decreased arousal and a decreased occurrence of SWD by increased arousal; (4) reproducibility with a predictable development; (5) a pharmacological profile that reflects the results obtained in clinical practice; (6) a unique developmental profile; and (7) potentiation of the SWD by drugs inducing petit mal-like seizures.

EEG characteristics of spike-and-wave discharges in GAERS

Cortical EEG

Several hundreds of adult female and male rats from this strain were implanted, under pentobarbital anaesthesia (40 mg/kg i.p.), with 4 single contact electrodes over the left and right frontoparietal cortex. The electrodes were screwed into the skull and soldered to a microconnector embedded in acrylic resin with anchoring screws on the skull. Four animals were fitted with 10 wire-electrodes distributed over the cortex.

For recording, the animals were placed in a plexiglass box, enclosed in a grounded metal screen. The rats were connected via flexible wires to the electroencephalograph and were freely moving under permanent observation. Light sensory stimulations were delivered to prevent sleeping.

The SWD started and ended abruptly on a normal background EEG. The mean frequency of spike-and-wave within a discharge was 8.8 ± 0.5 cps

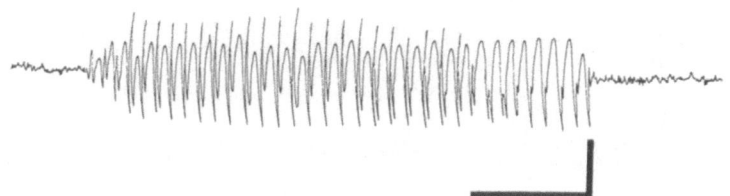

Fig. 1. Spontaneous cortical spike-and-wave discharge recorded in a GAERS. Calibration 1 s, 400 µV

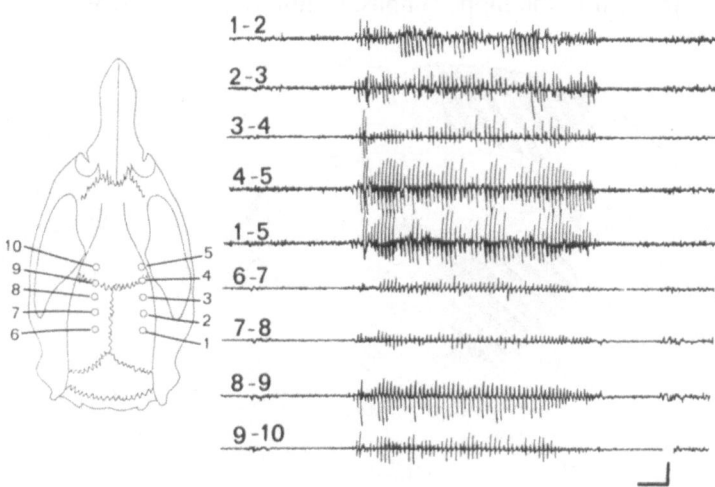

Fig. 2. Spontaneous spike-and-wave discharge recorded with ten single-contact cortical electrodes. The spike-and-waves predominate in the more anterior and lateral sites. Calibration 1 s, 400 µV

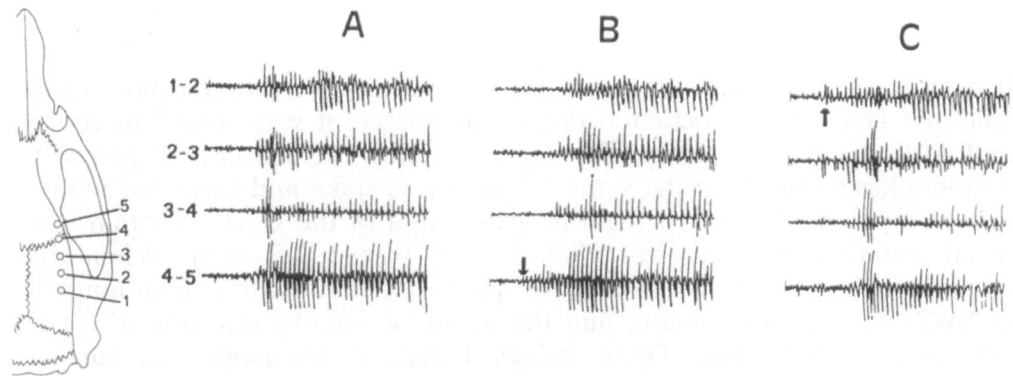

Fig. 3. Three different spike-and-wave discharges recorded in the same rat, with five single contact cortical electrodes: **A** SWD appears simultaneously all over the cortex; **B** SWD starts in the frontoparietal cortex (electrode 5); **C** SWD starts in the occipital cortex (electrodes 1–2). Calibration 1 s, 400 µV

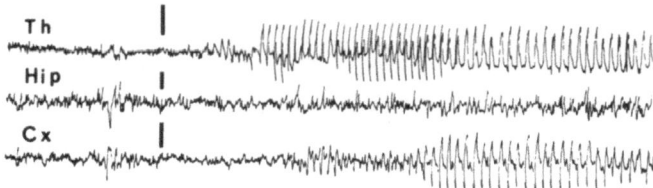

Fig. 4. Simultaneous recordings from the cortex (Cx), dorsal hippocampus (Hip), and lateral posterior thalamic nuclei (Th). The SWD starts in the thalamus and is not apparent in the hippocampus. Calibration 1 s, 200 μV

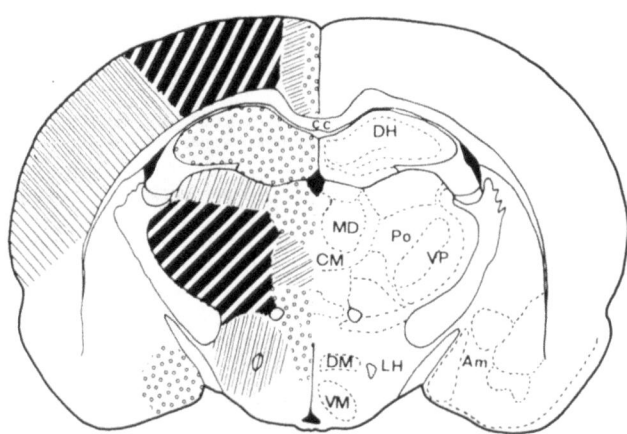

Fig. 5. Schematic mapping of SWD on a coronal section of a rat brain (according to the atlas of Paxinos and Watson, 1982). *Am* amygdala; *DH* dorsal hippocampus; *LH* lateral hypothalamus; *DM* dorsomedial nucleus of the hypothalamus; *VM* ventromedial nucleus of the hypothalamus; *CM* centromedial, *MD* mediodorsal, *Po* posterior, *VP* ventroposterior nucleus of the thalamus. Thick stripes, sustained and large SWD; thin stripes, small and irregular SWD; dots, no SWD recorded

(range 7–11 cps). The voltage of the spike-and-wave varied from 300 to 1,000 μV and usually fluctuated during discharges; it was always three- to tenfold over baseline activity (Fig. 1). Sometimes, the discharge began with a monophasic activity at the same frequency as spike-and-wave but with a lower amplitude. In rats of the 10th generation of the GAERS strain, the mean duration of SWD was 17.2 ± 9.8 s (range 0.5–75 s). When the animals were maintained in a state of quiet wakefulness, the mean number of SWD was 1.3 per minute, and the mean cumulative duration of SWD was 24.3 ± 7.5 s/min. These values remained unchanged in further generations.

In rats with 10 cortical single contact electrodes, SWD predominated in the most lateral (3 mm lateral to the midline) and frontoparietal sites (Fig. 2). The initiation of SWD might vary from one discharge to another in the same rat. Usually, SWD appeared simultaneously all over the cortex

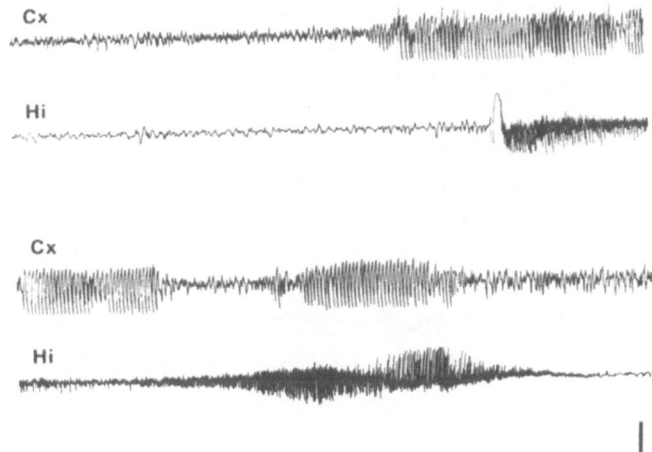

Fig. 6. Simultaneous recording of the cortical (Cx) reference EEG, and of the hippocampus (Hi). When the bipolar electrode is lowered into the hippocampus, an artefact is seen (upper Hi trace) followed by a focal subclinical hippocampal seizure. Concomitant spontaneous cortical SWD occur independently. Calibration 1 s, 400 µV

(Fig. 3A). Sometimes they started in the frontoparietal regions (Fig. 3B) and rarely in the occipital cortex (Fig. 3C).

Depth EEG

In order to determine which brain structures are involved in SWD, chronic and acute EEG recordings were performed with depth bipolar electrodes. In chronic experiments, the EEG was recorded simultaneously in the cortex, the thalamus and the hippocampus (Vergnes et al., 1987). In acute experiments, the recordings were made from movable bipolar electrodes in curarized rats. Fixed cortical electrodes were used as an EEG control before and during the experiment (Vergnes et al., 1990a,b).

The largest SWD were recorded from the frontoparietal cortex and the posterolateral thalamus. In some rats, the beginning of the lateral thalamic SWD occasionally preceded that of the cortical SWD (Fig. 4). Small-amplitude or delayed SWD were present in the striatum, lateral hypothalamus and ventral tegmentum. SWD were absent or considerably reduced in the anterior and midline nuclei of the thalamus. No SWD were recorded from the limbic structures: hippocampus, septum, amygdala, cingular and pyriform cortex (Figs. 4, 5). In acute experiments, penetration of the movable electrode into the hippocampus sometimes elicited focal, high-amplitude, subclinical EEG seizures which were independent of cortical SWD. The two types of seizure might be recorded simultaneously, i.e., SWD in the cortex and partial limbic seizures in the hippocampus (Fig. 6). Similar discharges involving the cortex and thalamus were also observed in other strains of rats with spontaneous SWD (Albe-Fessard and

42 C. Marescaux et al.

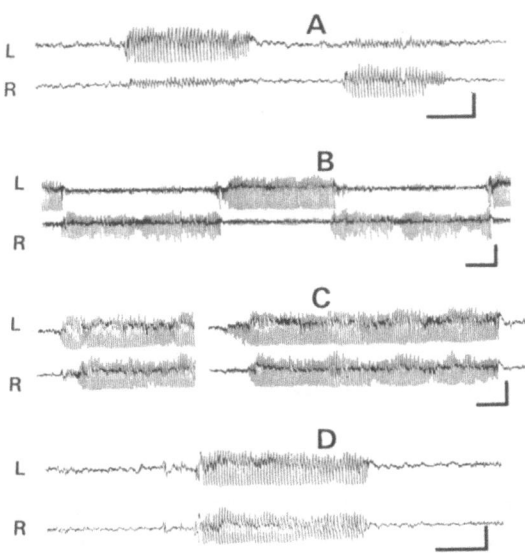

Fig. 7. Different patterns of SWD recorded in the left (L) and right (R) frontoparietal cortex of a same rat with a full transection of corpus callosum. **A** unilateral SWD; **B** SWD alternating abruptly from one hemisphere to the other; **C** bilateral SWD starting asynchronously; **D** bilateral and synchronous SWD. Calibration 2 s, 400 μV

Lombard, 1963; Klingberg and Pickenhain, 1968; Semba et al., 1980; Chocholová, 1983; Semba and Komisaruk, 1984; Buzsáki, 1990a,b) and in cats with SWD inducted by penicillin (Avoli and Gloor, 1982; Avoli et al., 1983).

Some authors reported SWD recorded from the dorsal hippocampus in other models of absence epilepsy in mice (Noebels, 1984) and in rats (Fariello and Golden, 1987; Serikawa et al., 1987). Differences between strains or artifacts due to monopolar recording in small animals may account for this discrepancy. In our experiments, as well as in others (Albe-Fessard and Lombard, 1983), no SWD were recorded with a bipolar electrode located within the hippocampus. Moreover, in GAERS, the occurrence of a focal seizure in the hippocampus does not interfere with generalized non-convulsive SWD recorded in the cortex, demonstrating that the two kinds of seizure function independently with distinct substrates.

Bilateral synchronization of SWD: role of the corpus callosum

In rats of the GAERS strain, the bilateral cortical and thalamic SWD are always synchronous. In 13 rats, the corpus callosum was transected with a surgical blade fixed in a stereotaxic apparatus and moved along the midline from 2 to 12 mm anteriorly to the lambda (4 to 4.5 mm dorsoventrally from the surface of the skull). After allowing about 7 days for recovery, bilateral electrodes were implanted on the cortex (10 rats), or simultaneously on the cortex and in the thalamus (3 rats).

In these rats, the bilateral synchronism of the cortical and thalamic SWD was modified (Vergnes et al., 1989). Three SWD patterns were seen (Fig. 7): (1) unilateral SWD occurred independently on each hemisphere, or sometimes alternated abruptly from one hemisphere to the other; (2) SWD started unilaterally and then continued bilaterally after variable delays (0.5 to several seconds); (3) bilateral synchronous SWD persisted in most animals, but this pattern was the least frequent (<10%). Thalamic SWD were always associated and synchronous with the homolateral cortical SWD, whatever the pattern of bilateral desynchronization. When the SWD were present simultaneously on both hemispheres, each single spike-and-wave complex was bilaterally synchronous.

In 4 rats, a lesion of the midline thalamic nuclei separating the thalamus in two parts did not affect SWD which occurred synchronously on both hemispheres. In 4 rats, a transection of the corpus callosum associated with a cut through the midline thalamus completely abolished, the bilateral synchronization of the SWD: SWD occurred independently on each hemisphere, and when they were simultaneous, the bilateral spike-and-waves were asynchronous (Vergnes et al., 1990b, this volume).

These results show that the corpus callosum plays a major role in the bilateral synchronization of SWD in GAERS. The corpus callosum is not the only structure involved, since its complete section does not totally abolish bilateral synchronous discharges. Synchronization may also develop through the midline thalamus, but this path is not necessary for synchronization to occur.

A similar desynchronization of bilateral SWD was produced by corpus callosum transection in the penicillin model in cats (Musgrave and Gloor, 1980).

Ontogenic development of SWD

Twenty-four rats from the sixth and seventh generations of the GAERS strain were fitted with cortical electrodes between the ages of 30 and 60 days. At 30 days, none of the 10 animals recorded had SWD. At 40 days, 4 out of 14 rats had SWD. The number of rats with SWD then increased regularly with age and reached 100% at 4 months. The first SWD were rare (1 or 2/hour) and short-lasting (1 to 3 s), with a low frequency (4 to 5 cps) and a variable morphology. With age, the number and duration of SWD increased and the morphology and frequency of spike and waves became more characteristic of adult SWD. The number of SWD reached its maximum (>1/min) around the age of 6 months. SWD could be recorded over months, and they never disappeared spontaneously. The duration of SWD increased with age and reached a maximum by 18 months (Marescaux et al., 1984a,d; Vergnes et al., 1986). A similar ontogenesis of spontaneous

SWD was described in other strains of rats (Chocholová, 1983; Coenen and Van Luijtelaar, 1987).

Behavioural characteristics of rats of the GAERS strain

Behaviour during spike-and-wave discharges

SWD were usually observed when the animals were motionless in a state of quiet wakefulness. During the discharges, the animals wore a fixed stare and were completely inert. Frequently rhythmic twitching of the vibrissae and of the facial muscles was observed (Vergnes et al., 1982; Marescaux et al., 1984a). Muscle tone in the neck was sometimes diminished, inducing a gradual and slight drop of the head (Vergnes et al., 1990b). At the end of the SWD, a sudden extension of the head preceded the recovery of the previous position. In some instances, SWD appeared when the rat was moving: the movement was suddenly interrupted and resumed as soon as the discharge stopped. During SWD, animals were unresponsive to non-relevant stimuli; however, the SWD were immediately interrupted by strong and unexpected sensory stimulation (Vergnes et al., 1982).

Relationship between spike-and-wave discharges and wakefulness

To analyse the relationship between SWD and wakefulness, 4 rats fitted with cortical, hippocampal, and myographic electrodes in the neck muscles were recorded for two 6-hour periods, one in the light and one in the dark phase of the 24-hour light-dark cycle (Lannes et al., 1988).

The total number of the duration of SWD were determined during the various vigilance states from the EEG and EMG data. Sixty-six per cent of the SWD started and then ended during the wakefulness EEG pattern. During the periods preceding sleep, the SWD were more frequent, and clearly longer, sometimes lasting more than one minute. They were seldom

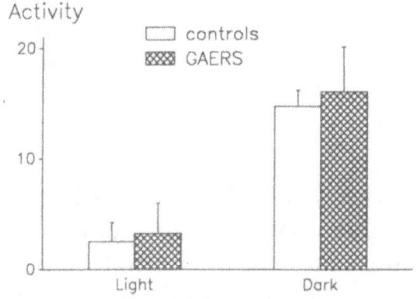

Fig. 8. Mean locomotor activity/h during the light and the dark phase (12/12 h) of the 24-h circadian period

seen during active behaviour. Twenty per cent of the SWD appeared during transition from wakefulness to slow-wave sleep, and 7% during transition from slow-wave sleep to arousal. Less than 7% started and ended in slow-wave sleep, and then usually during the first minute of a sleep episode. SWD were exceptional in paradoxical sleep (0.2%).

Similar relationships between spontaneous SWD and vigilance were described for different strains of rodents (Vanderwolf, 1975; Radil et al., 1982; Kaplan, 1985; Van Luijtelaar and Coenen, 1986).

Behavioural abilities of rats from the GAERS strain

An analysis of behavioural parameters was undertaken to reveal possible differences in the abilities of adult male rats of the GAERS strain and age-matched males of the seizure-free control strain. The animals were placed in individual cages for the duration of the experiments. When necessary, the EEG was recorded through permanently implanted cortical electrodes. Behavioural results were compared using the non-parametric test of Mann-Whitney (for details see Vergnes et al., 1991).

Spontaneous circadian locomotion was measured in the rat's home cage by the number of interruptions per hour of two infrared light beams (Fig. 8). No difference appeared in the usual 24-h rhythmicity between the GAERS (n = 14) and the controls (n = 14).

Exploratory activity in a non-familiar environment was measured when animals were placed for 6 min into a circular (diameter 1 m), lighted, open-field. Locomotion and rearing did not differ between the GAERS (n = 10) and the controls (n = 10).

Interspecific aggression was assessed by placing a mouse into the rat's cage for 1 hour; 5/10 GAERS and 5/10 controls killed the mouse.

Social interactions with a non-familiar male conspecific were recorded on video-tape during 8 min sessions. Different behavioural items were encoded using a microcomputer and processed in terms of cumulated duration per session (Depaulis, 1983). No difference either in non-social activities (exploration, grooming) or in social interactions (partner-investigation and allogrooming, offensive or defensive behaviour) appeared between GAERS (n = 10) and controls (n = 10), whether the animals were in a situation of residents in their home cage, or were introduced as intruders into the partner's cage.

Avoidance-learning trials were performed in a two-way shuttle box. Through the grid floor, an electric scrambled 0.5 mA shock was delivered, heralded by a 9 s acoustic signal. The rat had to learn to avoid the shock by crossing into the opposite compartment while the sound was still audible. When no avoidance occurred, the acoustic signal and the shock were prolonged for further 11 s, during which escape to the safe compartment was possible. After a 10-s rest, the next trial was started in the opposite compartment (10 trials per session, 12 sessions in 2 days). Avoidance or

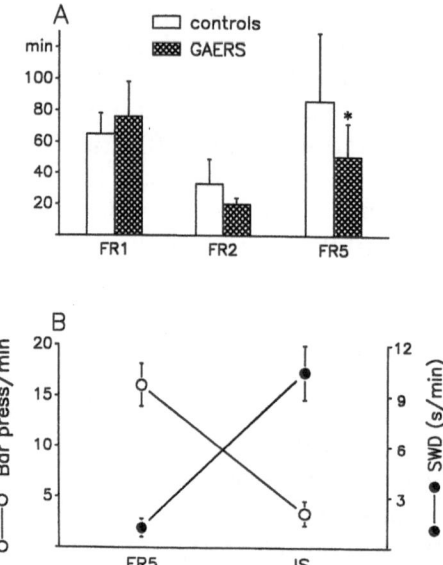

Fig. 9. A Learning in a Skinner box: time spent to reach the criterion of reward in FR1, FR2 and FR5 schedules. *p < 0.05. **B** Mean number of bar-presses and duration of SWD per min in GAERS during FR5 sessions and non-reinforced intersession (IS)

escape was considered to have been learned when it occurred in more than 8 trials per session. During the last three sessions, 3/10 GAERS had learned to avoid and 5 others to escape the shock. Among the controls, only 4/10 had learned to escape and no one avoided the shock.

Instrumental learning with a fixed-ratio food reinforcement was performed in a Skinner box. The animals, maintained at 85% of their normal body-weight, were trained to press a bar to obtain a 45-mg food pellet. The time necessary to obtain successively 25 pellets in a "fixed reinforcement 1" schedule (FR1), 25 pellets in a FR2 and 50 pellets in a FR5 schedule, was measured. The GAERS (n = 10) and the controls (n = 10) did not differ in ability to learn how to obtain food on a FR1 and FR2 schedule. When 5 bar presses were required to obtain food (FR5), the GAERS worked faster than the controls (Fig. 9A).

After training, the EEG was recorded in GAERS for three 20-min FR5 sessions, interrupted by two 10 min-periods without reinforcement, signaled by switching off the light. During the FR5 sessions with reward, SWD were suppressed as long as the rats were working for food. During the last session, when satiation begins, bar-pressing was occasionally discontinued and some SWD appeared. During the non-reinforced intersessions, bar-pressing ceased and many SWD were recorded (Fig. 9B).

Instrumental conditioning using an association of an acoustic signal with bar-pressing for food, was trained in 5 GAERS, maintained at 85% of their body-weight and under continuous EEG control. The rats had to respond within 10 s to a mild sound (0.8 s) to obtain a food pellet (randomly 2/min). When trained, the animals worked for food over 2 hours. No SWD were recorded during this time. When satiation occurred, bar-pressing in response to the sound became irregular and finally ceased. SWD then appeared and usually were not interrupted by the sound. No response was ever recorded during a SWD. When the same animals were placed in an empty cage while food-deprived, many SWD were recorded. When food was given, feeding started immediately and SWD disappeared.

These results show that rats with absence epilepsy are not impaired in usual behaviours such as spontaneous activity, exploration, feeding, social interactions or learning of positively or negatively reinforced tasks. Sexual and reproductive behaviours also appear normal.

EEG recording during the performance of various tasks showed that no SWD discharge occurs as long as the animal is motivated to obtain the reinforcement. SWD supervene as soon as the animal becomes inactive, either because the reward is withdrawn, or because the motivation to get it has vanished. Consequently, SWD clearly appear not to interfere with active behaviours, as they only occur when attention and activity are

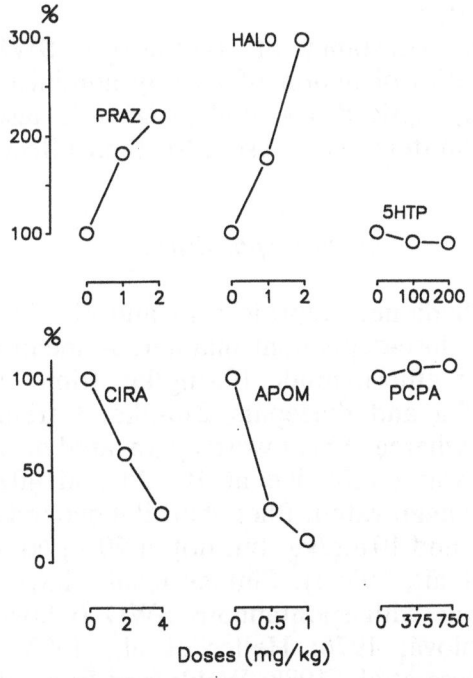

Fig. 10. Effects of prazosin (PRAZ, α1 antagonist), cirazoline (CIRA, α1 agonist), haloperidol (HALO mixed D1–D2 antagonist), apomorphine (APOM, mixed D1–D2 agonist), 5 hydroxytryptophan (5-HTP, 5-HT precursor) and parachlorophenylalanine (PCPA, inhibitor of 5-HT synthesis) in GAERS. 100% represents the mean cumulative duration of SWD in s/min during the period before treatment. The effects of drugs are related to this reference period. Abscissae: dose in mg/kg i.p

already suppressed. During SWD, responsiveness to mild or non-significant stimuli is then abolished.

Pharmacological characteristics of spike-and-wave discharges in GAERS

Each experiment was performed on a group of 6 to 8 rats fitted with 4 single contact electrodes. After the freely moving animals were adapted to the registration environment for 15 min, a reference EEG has recorded for 20 min. The drug was then injected intraperitoneally (i.p.), and the EEG was recorded for the following 60 to 120 min, according to the duration of effect of the given drug. During the recording periods, the rats were continuously watched and prevented from falling asleep by gentle sensory stimulation. Control experiments were performed in which the solvent was injected. EEG recordings were analysed over successive 20-min periods. For each group of rats, the mean cumulative duration, in seconds per 20 min, was calculated during the reference and the post-injection periods. The post-injection periods were compared to the reference period using a non-parametric analysis of variance for related samples (Friedman test). Responses to each dose versus the control (dose 0) were then compared using the Wilcoxon test.

The frequency and constancy of spontaneous SWD permit a clear evaluation of the kinetics of action of acutely administered drugs. Three types of drugs, anti-epileptic drugs, epileptogenic drugs and drugs interfering with neurotransmitters, were tested for their effects on SWD.

Antiepileptic drugs

Ethosuximide, trimethadione, valproic acid and classical benzodiazepines suppressed SWD in a dose-dependent manner. A mean efficacy exceeding 90% was observed for ethosuximide 100 mg/kg, trimethadione 200 mg/kg, valproic acid 200 mg/kg and diazepam 2 mg/kg. Carbamazepine was ineffective at 10 mg/kg, whereas SWD were aggravated at doses of 20, 40 and 80 mg/kg. Phenytoin was ineffective at 10, 20 and 40 mg/kg; at 80 and 160 mg/kg, SWD were aggravated. Phenobarbital evoked biphasic effects: it was effective at 2.5, 5 and 10 mg/kg, but not at 20 mg/kg (Marescaux et al., 1984a,c; Micheletti et al., 1985b). Similar results have been obtained in other strains of rodents with spontaneous SWD (Chocholová and Radil-Weiss, 1973; Chocholová, 1976; Heller et al., 1983; Kleinlogel, 1985; Peeters et al., 1988; Sasa et al., 1988; Wahle and Frey, 1990).

New potentially anti-epileptics drugs were also tested. Gamma-vinyl-GABA aggravated SWD and progabide was ineffective; 2-en-valproate and partial agonists of benzodiazepine receptors suppressed SWD (Jensen et al., 1984; Löscher et al., 1984; Vergnes et al., 1984; Marescaux et al., 1987a; Bernasconi et al., 1988).

Epileptogenic drugs

The SWD were increased by drugs commonly used to induce petit mal-like seizures: pentylenetetrazol at low doses (10–20 mg/kg); gamma-hydroxy-butyrate (250–500 mg/kg); THIP (4–12 mg/kg) and penicillin (1.25–2.5 × 10^6 UI/kg) (Marescaux et al., 1984a, 1987b; Depaulis et al., 1988a; Vergnes et al., 1984, 1985, 1990b; Snead, this volume).

Drugs interacting with neurotransmitters

Drugs interacting with noradrenergic (NA) and dopaminergic (DA) neurotransmissions

Drugs which reduce αNA transmission, such as antagonists of the α1 post-synaptic receptor (prazosin 0.5–2 mg/kg), or agonists of the α2 presynaptic receptor (clonidine 0.01–0.04 mg/kg), increased the SWD. The drugs that increase αNA transmission, agonists of the α1 postsynaptic receptor (cirazoline 1–4 mg/kg) or antagonists of the α2 presynaptic receptor (yohimbine 0.5–4 mg/kg) reduced the SWD. The βNA-receptor agonist isoprenaline (25–100 mg/kg) and its antagonist propranolol (5–20 mg/kg), did not change the SWD (Fig. 10).

The mixed D1–D2 antagonists (haloperidol 0.5–3 mg/kg, and other neuroleptics) induced an increase in the duration of SWD. The mixed agonists simultaneously activating the D1 and D2 receptor (L-Dopa 50–200 mg/kg, apomorphine 0.1–1 mg.kg) induced a dose-dependent suppression of SWD. Drugs interacting specifically with only one type of DA receptor had less effect on SWD. The specific antagonists of D2 (sulpiride 25–100 mg/kg, etc . . .) and the specific agonists of D2 (lisuride 0.125–0.5 mg/kg, etc . . .) left the SWD unchanged (Fig. 10).

Similar results were obtained in cats with penicillin-induced generalized epilepsy (Quesney and Reader, 1984) and in several models of generalized non-convulsive seizures in rats: other strains with spontaneous SWD (Kleinlogel, 1985; Buzsáki et al., 1990a,b; Frey and Voits, 1991), flash-evoked after-discharges (King and Burnham, 1980, 1982), seizures induced by low doses of pentylenetetrazol (Libet et al., 1977). The role of NA in mice varies from one strain to another. In tottering mice with a genetic pro-liferation of NA nerve terminals, the spontaneous SWD are reduced after depletion of catecholamines (Noebels, 1984). By contrast, in DBA/2 mice, the spontaneous SWD are aggravated by the αNA antagonists as well as by the βNA antagonists (Ryan, 1985b).

Thus DA and NA participate in the control of SWD in GAERS (Micheletti et al., 1987; Warter et al., 1988, 1990) (Fig. 10). However, a deficiency in NA and DA neurotransmission is not likely to be involved in the generation of spontaneous SWD, since SWD could not be induced by any of the DA or NA antagonists in rats from the SWD-free control strain.

50 C. Marescaux et al.

Similarly, bilateral lesions of the locus coeruleus (Lannes et al., 1991) or the substantia nigra (unpublished data), which transiently aggravated the spontaneous SWD, never induced SWD in control rats.

Drugs interacting with serotoninergic (5HT) neurotransmission

The 5-HT precursor 5-hydroxytryptophan (50–200 mg/kg), induced no change in the SWD. The inhibitors of 5-HT reuptake had variable effects, most probably due to a concomitant effect on catecholamine uptake: at high doses, fluvoxamine ($\geq$100 mg/kg) produced a slight reduction of SWD, whereas indalpine ($\geq$12.5 mg/kg) increased them. Parachlorophenylalanine (375–750 mg/kg), an inhibitor of 5-HT synthesis, methysergide (12.5–50 mg/kg), a non-selective antagonist of the 5-HT receptor, as well as ritanserin (20–80 mg/kg), a specific antagonist of 5-HT$_2$ receptors, did not affect SWD (Fig. 10).

Thus, 5-HT appears not to intervene in the control of SWD in GAERS (Marescaux et al., 1992a). The role of 5-HT agonists and antagonists was not studied in other models of SWD. Dusser and Peroutka (1990) showed that 5-HT receptors are normal in tottering mice with SWD.

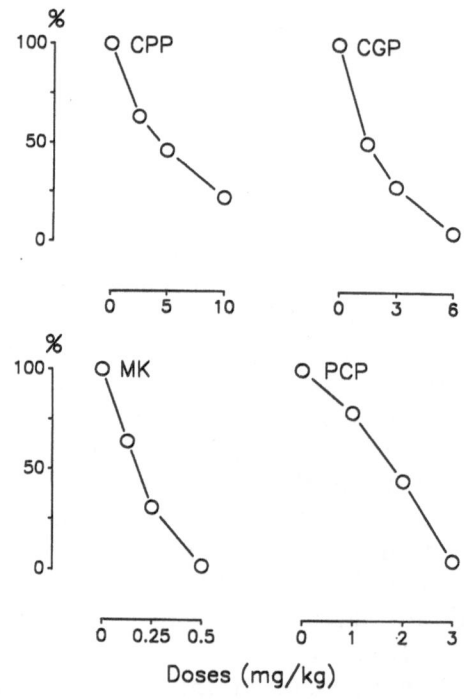

Fig. 11. Effects of NMDA antagonists in GAERS. CPP and CGP 40 116 (CGP): competitive antagonists; MK 801 (MK) and PCP: non-competitive antagonists. Conventions as in Fig. 10

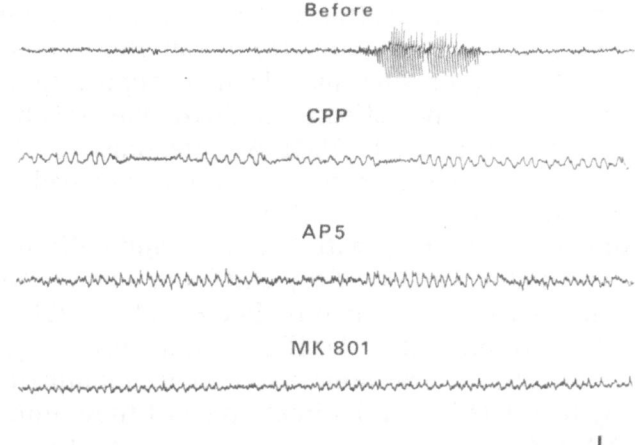

Fig. 12. Effects of NMDA antagonists on EEG activity (CPP 10 mg/kg, AP5 400 mg/kg, MK 801 0.5 mg/kg). Calibration 2 s, 400 µV

Drugs interacting with N-methyl-D aspartic (NMDA) neurotransmission

Systemic administration (20–100 mg/kg i.p.) of NMDA tended to decrease SWD in GAERS. At doses exceeding 75 mg/kg, some animals had convulsions. Because of the poor penetration of NMDA through the blood-brain barrier, NMDA was also administered intracerebroventricularly (i.c.v.). As with i.p. injections, i.c.v. NMDA never aggravated the SWD, but induced a dose-dependent decrease. At a dose of 40 ng in 5 µl, NMDA completely suppressed the SWD for more than 120 min. At higher doses (≥1 µg in 5 µl), NMDA produced convulsions within a few minutes (results not shown).

Five competitive antagonists of NMDA: AP5, AP7, CPP, CGP 43487, CGP 40116 and five non-competitive antagonists: SKF 10047, MK 801, ketamine, phencyclidine (PCP) and its thienyl-derivative (TCP) were used. The specific and non-specific antagonists of NMDA all suppressed the SWD (Fig. 11). Major behavioural and EEG side-effects were induced by these drugs: continuous sniffing up and down, horizontal movements of the head, rotations, ataxia associated with slow and paroxysmal high-amplitude waves on the EEG (Fig. 12).

Thus, drugs that interact with the NMDA receptor interfere with SWD (Marescaux et al., 1992a). The role of NMDA in the control of SWD and of partial and convulsive seizures appears, however, to depend on different mechanisms. Suppression of SWD by NMDA antagonists is obtained with high doses only (MK 801 > 0.25 mg/kg; PCP > 2 mg/kg; CPP > 5 mg/kg; CGP 40116 > 3 mg/kg) (Fig. 11), when side-effects are induced, whereas convulsive seizures are suppressed by lower doses (Patel et al., 1988; Marescaux et al., 1992a). NMDA administration did not aggravate SWD, but decreased them, and never induced SWD in control rats at any dose (unpublished data). These data suggest that SWD do not simply result

52   C. Marescaux et al.

from an excessive NMDA neurotransmission. Glutamatergic synapses are
involved in the thalamocortical circuitry underlying SWD. Their dysre-
gulation at NMDA-receptor sites may lead to suppression of the SWD.
Alternatively, the suppressive effects obtained may result from a modification of the level of arousal: NMDA antagonists and NMDA itself
provoke motor activation and increased arousal respectively, states that are
incompatible with the appearance of SWD.

Similar results were observed with ketamine and CPP in an other strain
of rats with spontaneous SWD by Frey and Voits (1991). Results differing
partially from our own were obtained by Peeters et al. (1989, 1990b) in rats
of the WAG/Rij strain: while NMDA antagonists suppressed SWD,
NMDA increased SWD several hours after i.c.v. administration. This late
aggravation may be related to an indirect effect of these injections, as i.c.v.
injections of NMDA at higher doses produce convulsions within a few
seconds or minutes.

Drugs interacting with gamma-hydroxybutyric (GHB) neurotransmission

The administration of GHB produced a dose-dependent increase in SWD.
At high doses (375 mg/kg), SWD became permanent with a reduced fre-

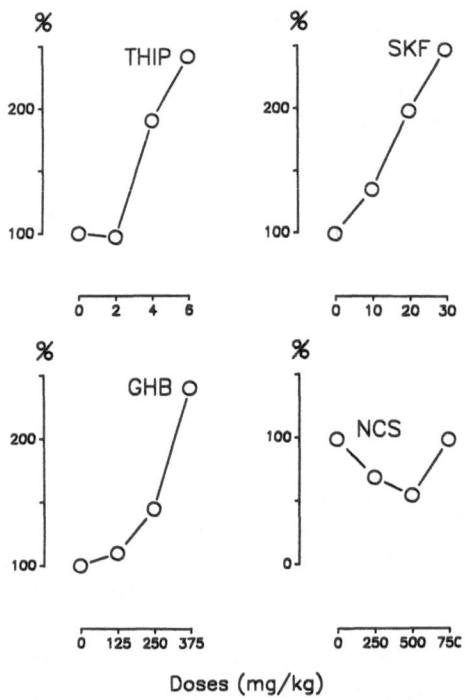

Doses (mg/kg)

Fig. 13. Effects of THIP (GABA$_A$ agonist), SKF 89 976 (SKF: inhibitor of GABA
reuptake), GHB and NCS 382 85 (NCS: GHB antagonist) in GAERS. Conventions as
in Fig. 10

quency of the spike and waves (4–5 c/s). NCS 38285, a specific antagonist of GHB receptor (Maître et al., 1990) has no effect on SWD at doses of 125, 250 and 750 mg/kg. At 500 mg/kg, it induced a moderate (about 50%) decrease (Fig. 13).

GHB is present in the mammalian brain and is derived from GABA. Specific binding sites of high and low affinity were found. GHB is now considered a potential neurotransmitter per se (Maître et al., 1990). In rodents, cats and monkeys, GHB administration produces an arrest of activity and myoclonic jerks simultaneous with SWD on the EEG. These seizures are responsive to anti-absence drugs. The GHB-induced seizures have been proposed as a model of generalized absence seizures (Snead, 1978, 1988, this volume).

GHB appears to control spontaneous SWD in rats of the GAERS strain (Depaulis et al., 1988a). In agreement with this hypothesis, an autoradiographic study has shown that [^{3}H]-GHB receptor binding was increased in the lateral thalamic nuclei of GAERS rats (Snead et al., 1990). However, the biochemical data (see Bernasconi et al., this volume) and the lack of effect of the GHB antagonist NCS 38285 after i.p. or intrathalamic injection (Liu et al., 1991a) do not support the hypothesis that excessive GHB neurotransmission is directly responsible for the genesis of SWD in GAERS.

Drugs interacting with gamma-aminobutyric (GABA) neurotransmission

The effects of GABA are mediated by $GABA_A$ (GABA/benzodiazepine complex) receptors and $GABA_B$ receptors. There are marked differences between $GABA_A$ and $GABA_B$ receptors in terms of pharmacological profile, mechanism of action and regional distributions. In this chapter, we report only on drugs interacting with the $GABA_A$ receptor. The effects of $GABA_B$ agonists and antagonists will be reported elsewhere in the same issue (Marescaux et al., this volume).

No dose of $GABA_A$ antagonists picrotoxin (0.5–2 mg/kg) or bicucullin (0.5–4 mg/kg) ever exerted a significant effect on the SWD. At high doses, convulsive seizures appeared (results not shown). All GABA-mimetics induced a dose-dependent increase in the duration of SWD: THIP (2–12 mg/kg), a $GABA_A$-receptor agonist, gamma-vinyl-GABA (GVG) (200–1,200 mg/kg), an inhibitor of GABA transaminase, and SKF 89976 (10–40 mg/kg), an inhibitor of GABA reuptake (Fig. 13). At high doses, the GABA-mimetics induced permanent SWD with a reduced frequency (5–6 c/s) or isolated spikes on a flat background (burst-suppression).

GABA-mimetics are inhibitors of seizures in many models of convulsive epilepsy. Conversely, they aggravate the seizures in all models of generalized non-convulsive epilepsy in rodents, as well as in cats (King, 1979; Fariello et al., 1980; Smith and Bierkamper, 1990). They also aggravate the bilateral spike and wave discharges induced by light flashes in the baboon

Papio papio (Meldrum and Horton, 1980). Moreover, administration of GABA-mimetics in non-epileptic animals induces bilateral SWD that resemble generalized non-convulsive seizures (Fariello and Golden, 1987).

GABA-mimetics facilitate the expression of SWD in rats of the GAERS strain and induce SWD in non-affected rats of the control strain (Vergnes et al., 1984, 1985; Micheletti et al., 1985a; Liu et al., 1990; Marescaux et al., 1992a). These results suggest that a GABAergic dysfunction may be involved in the genesis of spontaneous SWD. However, $GABA_A$ antagonists are unable to suppress SWD and no difference in $GABA_A$ receptors have been found between GAERS and rats of the control strain (Knight and Bowery, this volume).

Recently, different groups of benzodiazepine (BZD) receptor ligands have been described. According to both electrophysiological and biochemical data, these compounds have been classified as agonists, inverse agonists or antagonists. (1) Injection of full agonists (diazepam, ZK 93423) suppressed SWD in a dose-dependent manner, and induced sedation with alteration of EEG background activity. (2) Injection of partial agonists (ZK 91296, Ro 166028, CGS 9896) suppressed SWD without sedation. (3) Injection of low doses of a full inverse agonist (DMCM) significantly increased the total duration of SWD; higher doses of DMCM induced convulsions. (4) Injection of partial inverse agonists (FG 7142, ZK 90886) aggravated SWD. (5) Injection of low doses of a benzodiazepine antagonist (Ro 151788) did not significantly modify SWD; however, this compound was able to reverse the effects of both agonists and inverse agonists. Thus, in our GAERS strain, the effects induced by the BZD receptor ligands constitute a continuum correlated with their activity at the receptor level.

Two lines of reasoning might explain the opposite effects of $GABA_A$ mimetics and BZD agonists in the GAERS strain. First, the anti-SWD effects of BZD agonists might depend on a selective activation of receptors localized in particular anatomical structures. Recently, we have shown that localized activation of $GABA_A$ receptors in the substantia nigra (Depaulis et al., 1988b, 1989, 1990a,b, this volume) or in the reticular nuclei of the thalamus (Liu et al., 1991b) suppressed SWD. Secondly, despite evidence for a GABAergic effect of BZD, their pharmacological efficacy may not be exclusively mediated by the $GABA_A$ receptor (Marescaux et al., 1985).

The sensitivity of rats of the GAERS strain to BZD receptor ligands and the fact that SWD can be induced in control animals as a result of administration of inverse agonists suggest a possible involvement of BZD receptors in the elicitation of spontaneous SWD (Jensen et al., 1984; Marescaux et al., 1984b, 1985, 1987a; Bernasconi et al., 1988). However, recent results obtained in GAERS failed to show any difference between GAERS and control animals, in both the affinity and the number of BZD receptors of the central type (Richards, personal communication).

Preliminary results obtained in our laboratory have shown that acetylcholine, adenosine and morphine are also involved in the control of SWD in GAERS, as they are in other models of absence epilepsy (Ryan, 1985a; Kostopoulos et al., 1987; Frey and Voits, 1991).

In conclusion, it appears that most neurotransmitters are involved in the elaboration or control of SWD. According to our results, generalized non-convulsive seizures in GAERS, in contrast to convulsive seizures, cannot be directly related to an excess of excitatory neurotransmission by NMDA, nor to a deficiency of GABAergic inhibition. The fact that $GABA_A$ and GHB agonists produce an increase in spontaneous SWD in GAERS and also generate spike-and-wave-like discharges in non-epileptic controls suggests that these neurotransmitters might be important in the generation of SWD in GAERS. Possible dysfunctions of these neurotransmissions are under investigation.

However, to demonstrate a causal relationship between a dysfunction in a given neurotransmission and generation of SWD, it would be necessary to show: (1) opposed effects of agonists and antagonists on SWD in GAERS; (2) induction of SWD in control rats by drugs aggravating the SWD in GAERS; (3) demonstration of differences in the amount and/or turnover of the neurotransmitter, or the number and/or affinity of its receptors. Up to now, no neurotransmitters have met these criteria.

Genetic transmission of spike-and-wave discharges in GAERS

In our initial colony of Wistar rats, 30% of the animals showed spontaneous SWD. Inbreeding of selected parents over a few generations produced a strain in which 100% of the rats were affected. Similarly, we selected a control strain free of SWD. These data demonstrated that SWD were genetically controlled.

We analysed the mode of inheritance of SWD in GAERS by performing a classical Mendelian cross-breeding study. Two hundred and seventy-five of the offsprings from different kinds of crossing were used: 101 rats from the 10th to 13th generation of the control strain; 40 rats from the 14th to 17th generation of the GAERS strain; 20 F1 hybrids from GAERS females × control males; 20 F1 hybrids from GAERS males × control females; 49 F2 (F1 × F1) offsprings; 25 back-crosses from F1 females × control males and 20 back-crosses from F1 males × control females.

At 4 months of age, the rats were fitted with standard EEG electrodes. EEG's were recorded for one hour every month, between 4 and 12 months (242 rats were still alive and recorded at 12 months). For each rat, the number and the mean-cumulative duration of SWD per minute was measured during each one-hour recording.

All 101 offsprings of parents from the control strain had a normal EEG without SWD, up to 11 months (Fig. 14, Na). At 12 months, 4 of the 81 remaining rats and at 24 months 1 of the 18 remaining rats (5%) showed short, low-amplitude irregular discharges (Fig. 14, Na).

All 40 offsprings of parents from the GAERS strain showed SWD on the first EEG at 4 months (Fig. 14,A). At 12 months, the mean cumulative duration of SWD was between 15 and 30 seconds per minute (Fig. 14A).

56 C. Marescaux et al.

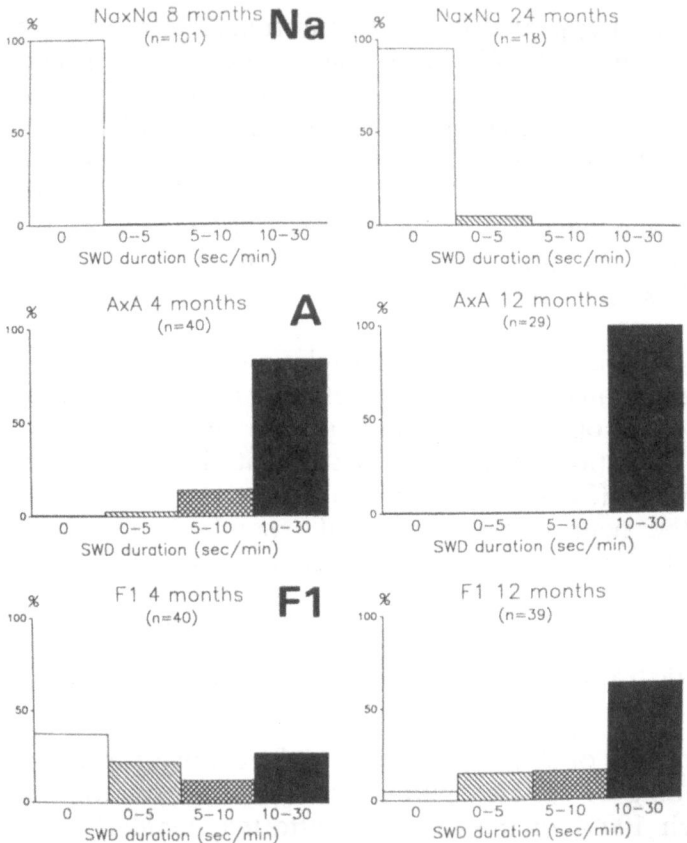

Fig. 14. Genetic transmission of SWD in Wistar rats. Percentage of rats showing SWD at different ages. **Na** non-absence (CONTROL) strain; **A** absence (GAERS) strain. **F1** offspring from CONTROL × GAERS reciprocal crosses. Non-affected rats: 0 s SWD/min. Affected rats are classified in 3 groups according to the mean cumulative duration of SWD per min

Twenty-five of the 40 (62%) F1 offsprings from control × GAERS reciprocal crosses showed SWD at 4 months, and 37/39 (95%) at 12 months. Both GAERS males × control females and GAERS females × control males produced a similar distribution of SWD among male and female offsprings. Inter-individual variability for age of appearance of SWD (4–12 months) and cumulative duration per minute (<2.5–25 sec) was extremely high (Fig. 14, F1).

SWD were recorded in 27/49 (55%) of F2 (F1 × F1) offsprings at 4 months, and in 38/44 (86%) at 12 months (Fig. 15A). As in F1, inter-individual variability for age of onset and duration was high. At 12 months, 70% of F2 from low-SWD parents (<2.5 sec SWD/min) and 100% of F2 from high-SWD parents (>10 sec SWD/min) were affected (Fig. 15B,C).

F1 were back-crossed with control rats. Seven of their 45 offsprings (16%) showed SWD at 4 months, and 27/40 (67%) at 12 months with high

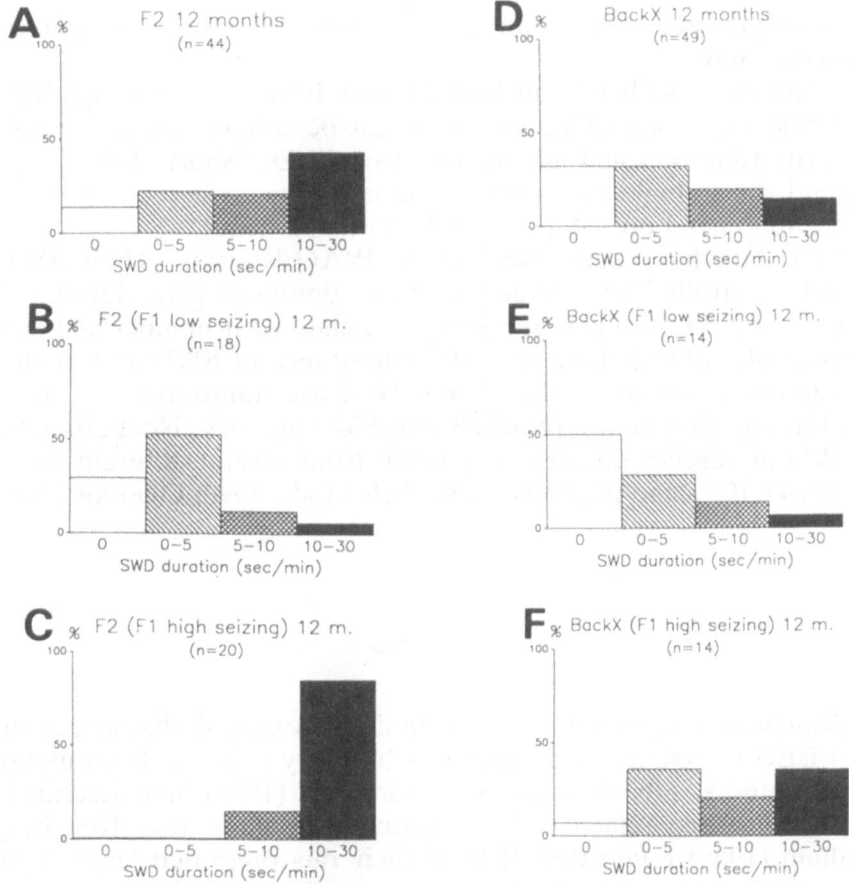

Fig. 15. Genetic transmission of SWD in Wistar rats. Percentage of rats showing SWD at 12 months. **A, B, C** F2: offspring from F1 × F1 crosses. **D, E, F** Back × (back-crosses): offspring from F1 × CONTROL reciprocal crosses. F1 low seizing: F1 parents displaying less than 2.5 s SWD/min. F1 high seizing: F1, parents displaying more than 10 s SWD/min

variability for duration (Fig. 15D). At 12 months, 50% of the back-crosses from low-SWD F1 (<2.5 sec SWD/min), and 93% of the back-crosses from high-SWD F1 (>10 sec SWD/min) showed SWD (Fig. 15E,F).

These data confirm that all GAERS showed SWD in their EEG, and that none of the control rats did so before 12 months of age. The 95% SWD rats in the F1 generation suggest that there is a dominant transmission. Similar SWD in males and females in the F1 generation indicate that the transmission is autosomal.

The high variability for duration and age of onset in F1, F2 and back-crosses suggests that the inheritance of SWD is probably not due to a single gene locus. Moreover, at 12 months, 86% and 67% of the rats showed SWD in the F2 and back-cross generations respectively. These scores are higher than the values one would expect if the occurrence were determined by one gene only.

Data obtained with F2 and back crosses from low- and high-SWD F1 suggest that the mode of transmission can be differentiated according to the severity (duration and age of onset) of SWD: "short" SWD appeared to depend on a single autosomal dominant gene, whereas "long" SWD appeared to depend on one or several additional genes.

Similar data have been obtained in WAG/Rij rats. Their SWD are apparently controlled by several genes: one dominant gene determines the occurrence of SWD, while other genes modulate their number and duration (Peeters et al., 1990a). However, the inheritance of SWD in rodents may vary from one strain to another. Thus SWD are transmitted in a recessive way in the tottering mouse (Noebels and Sidman, 1979; Noebels, 1984). In fact, SWD in absence epilepsy may result from several different mutations that provoke the same phenotype (Noebels et al., 1990; Qiao and Noebels, 1991).

Discussion

The spontaneous occurrence of rhythmic paroxysmal discharges on the cortical EEG of rodents, especially in laboratory rats, has been mentioned by many authors. Libouban and Oswaldo-Cruz (1958) first described such patterns, which they related to facial twitching in albino rats. Klingberg and Pickenhain (1968) found that 20% of their rats presented large "spindle" discharges (7 to 10 cps) in the frontal cortex. These discharges occurred in awake, but quiet animals and could be elicited by an electric shock applied to the thalamus (Kohler and Klingberg, 1969). Since 1970, Chocholová and her group have worked on this EEG pattern, which appeared in 80% of chronically implanted rats (Chocholová and Radil Weiss, 1970, 1973; Chocholová, 1976, 1983). Vanderwolf has observed the same paroxysmal pattern in the neocortex of some rats (Whishaw and Vanderwolf, 1971; Vanderwolf, 1975; Vanderwolf and Robinson, 1981).

Since 1975, more than 100 papers have mentioned, in various rodents, similar paroxysmal discharges, which could be observed in 10 to 90% animals of various laboratory colonies: guinea-pigs (Hammond et al., 1979), mice (Kaplan et al., 1979; Noebels and Sidman, 1979; Noebels et al., 1990; Ryan and Sharpless, 1979; Ryan, 1984) and rats (for review, see Vergnes et al., 1990b).

Four hypotheses have been considered. According to different authors, these EEG patterns might be: (1) an EEG artefact; (2) a physiological phenomenon; (3) a lesional epileptic phenomenon; (4) a spontaneous, genetic epileptic phenomenon.

The cortical recording of an artefact resulting from facial movements (Libouban and Oswaldo-Cruz, 1958) must be ruled out. Some rats remain completely motionless during the discharges, and bursts often occur in the absence of clonic twitches of the vibrissae or of the facial muscles. Moreover, SWD are recorded in curarized rats. Conversely, voluntary or passive movements of the head and vibrissae never produce these EEG patterns.

Many authors consider these SWD to be a physiological event that is characteristic of rodents. These bursts were related to behaviour as they occur selectively during certain behavioural states (Vanderwolf, 1975; Vanderwolf and Robinson, 1981). Various interpretations have been suggested. These include spindles occurring in a particular sleeping state (Timo-Iaria et al., 1970); activity corresponding to the "rythmes de veille immobile" or the sensorimotor rhythms of cats and primates (Radil et al., 1982; Chocholová, 1983); activity similar to human alpha or mu rhythms (Semba et al., 1980; Semba and Komisaruk, 1984); age-dependent spontaneous EEG bursts (Aldino et al., 1985; Aporti et al., 1986) or "high-voltage spindles" (Buzsáki et al., 1988a,b). The similar morphology of SWD and human EEG patterns during absence seizures has been noted (King, 1979), but the absence of behavioural convulsions led the authors to reject the hypothesis of epileptic activity (Klingberg and Pickenhain, 1968; Vanderwolf, 1975; Hammond, 1979). Thus, Kaplan (1985) concluded in a review of this question that the epileptic nature of rodent electrocortical poly-spiking is still unproved and that there is strong evidence for the proposal that such spiking is a normal EEG pattern.

Many authors, however, believe that these SWD are clearly pathological and appear as a result of such manipulations as cerebellectomy (Dow et al., 1962), cortical application of iron (Willmore et al., 1978), cortical or subcortical application of cobalt (Chocholová and Radil-Weiss, 1970; Roldan et al., 1970), 6-hydroxydopamine lesions of the substantia nigra (Buonamici et al., 1986), thalamic deafferentation by dorsal root sections (Albe-Fessard and Lombard, 1983), and slowly developing disease of the central nervous system caused by the spread of scrapie agents (Bassant et al., 1984, 1987). However, in most of these experiments, spontaneous SWD were also mentioned in control animals.

When these discharges are recorded in untreated animals, neither the effect of anaesthesia, nor a possible cortical damage produced by implantation material could reasonably be considered as the source of these patterns. Thus, for several authors, these EEG patterns in the rodents are spontaneous, even if they can be accentuated by various kinds of lesions. They consider this paroxysmal EEG pattern as a spontaneous model of centrencephalic or non-convulsive, or petit mal epilepsy in tottering mice (Noebels and Sidman, 1979) and in rats (Robinson and Gilmore, 1980). We demonstrated the validity of the SWD in rats as a model of petit mal epilepsy by analysing their symptomatic, pharmacological and aetiological analogies with human absences and other models of petit mal epilepsy (Vergnes et al., 1982, 1990b; Marescaux et al., 1984a,c,d, 1992a). Our results have been replicated by others, who drew similar conclusions

(Van Luijtelaar and Coenen, 1986; Serikawa et al., 1987; Buzsáki et al., 1990a,b).

Electroclinical semeiology of seizures in human absence epilepsy and of SWD in genetic absence epilepsy in rats from Strasbourg

Human absence seizures start and end abruptly, and may be associated with bilateral clonus of the eyes and the neck muscles (Loiseau and Cohadon, 1970; Berkovic et al., 1987). Absence seizures are classically concomitant with unresponsiveness to environmental stimuli and cessation of activity (Mirsky et al., 1986). The behavioural symptomatology of GAERS during a SWD is very close to that of absences. Despite SWD, the ability to perform spontaneous and learned behavioural activities is unimpaired in GAERS. Similarly, in children with typical absence epilepsy, intelligence is considered to be normal (Lennox and Lennox, 1960; Aicardi, 1986).

Petit mal absences may occur as frequently as several hundred times per day, mainly during quiet wakefulness, inattention, and in the transitions between sleep and waking — that is, drowsiness, slipping into slow-wave sleep, and nocturnal and morning arousal. They are interrupted by attention and unexpected sensory stimulation (Jung, 1962; Guey et al., 1969; Loiseau and Cohadon, 1970; Billiard, 1982). The relation of rat SWD with vigilance states is similar.

The morphology of the rat SWD is very similar to that of human spike-and-waves during an absence seizure. However spontaneous SWD in humans and rats differ in two ways: the frequency of the spike-and-waves and the age of seizure development. The spike-and-wave frequency in human absences is classically 3 cps, whereas in our rat model the frequency varies from 7 to 11 cps. In rodents, it is, in fact, impossible to elicit generalized non-convulsive seizures at a frequency around 3 cps (MacQueen and Woodbury, 1975; Avoli, 1980). In penicillin-induced seizures in cats, the mean frequency is 4.5 cps (Prince and Farrell, 1969). Only in primates can 3 cps SWD be elicited (Snead, 1978). The frequency of spike-and-waves in generalized non-convulsive seizures seems to be dependent on brain development and size for each species.

In humans, absence epilepsy is a disease of childhood, which tends to disappear with adulthood, although a form of absence also appears at puberty (Loiseau and Cohadon, 1970). In rats, the SWD appear after full maturation of cortical electrogenesis by 4 to 5 weeks of age for spontaneous SWD, at 3 weeks for pentylenetetrazol-induced SWD (Schickerova et al., 1984), and at 4 weeks for gammahydroxybutyrate-induced SWD (Snead, 1988). The spontaneous seizures persist until death. Since the process and degree of maturation of the human and the rat brain differ profoundly, it is not surprising that the ontogenetic development of petit mal epilepsy is quite different in the two species.

Pharmacological reactivity of absence epilepsy in humans and in rats

Rat SWD are suppressed by 4 main anti-epileptics (ethosuximide, trimethadione, valproic acid, benzodiazepines), which are also the only anti-epileptics effective against human absences (Loiseau and Cohadon, 1970; Marescaux et al., 1984a,d). Carbamazepine and phenytoin are as ineffective at low doses and aggravating at high doses for rat SWD as in human absences (Loiseau and Cohadon, 1970; Marescaux et al., 1984a,d). Discrepancies in the efficacy of phenobarbital in patients with petit mal epilepsy (Micheletti et al., 1985b) may be related to differential effects observed on rat SWD, according to the dose used. Finally, progabide is ineffective against rat SWD and gamma-vinyl-GABA aggravates rat SWD; and these two new antiepileptics were recently shown to produce the same effects in human petit mal epilepsy (Stefan et al., 1988; Luna et al., 1989). The SWD in rats are also increased by the seizure-inducing drugs that are commonly used as models of petit mal: GABA-mimetics, gamma-hydroxy-butyrate, pentylenetetrazol and penicillin (Fariello and Golden, 1987; Snead, 1988, this volume).

Pathophysiology of absence epilepsy in humans and in rats

No structural lesion of any kind — anatomical or biochemical — has ever been identified as the substrate of absence epilepsy (Berkovic et al., 1987; Gloor and Fariello, 1988). Its cause is regarded as genetic. In monozygotic twins, concordance rates of 84% for EEG discharges, and of 75% for absence seizures were found, while dizygotic twins showed no concordance (Lennox and Lennox, 1960). The high incidence of the presence of 3 cps spike-and-waves in first-degree relatives best fits an irregular autosomal dominant mode of inheritance, the gene having its highest penetrance in childhood and early adolescence (Metrakos and Metrakos, 1974; Gloor et al., 1982). Alternatively, several genetic factors have been suggested as being responsible for petit mal epilepsy (Doose et al., 1973).

There are no gross histological abnormalities in GAERS (Mouritzen Dam et al., 1989). Rats' SWD are genetically transmitted. The first results of a genetic analysis suggest an autosomal, dominant transmission. The variable expression may be due to the existence of several different genes controlling the number and duration of SWD.

The feline penicillin model of generalized absence seizures seems to be caused by an abnormal oscillatory pattern of discharges that involve a thalamocortical loop (Avoli and Gloor, 1982; Avoli et al., 1983; Gloor and Fariello, 1988). These thalamocortical circuits sustain the physiological spindles (Steriade and Deschenes, 1984) that evolve into SWD upon peni-cillin administration. Although this transition appears to occur first in the cortex, neither the cortex nor the thalamus can sustain the spike-and-wave pattern alone (MacLachlan et al., 1984a,b; Gloor and Fariello, 1988).

In humans, EEG recordings obtained simultaneously in the cortex and thalamus during absence seizures show that the rhythmic activity is synchronous in both structures and sometimes appears first in the thalamus (Williams, 1953). Similar observations in rats (Vergnes et al., 1987, 1990a,b, this volume) confirm the dependence of cortical and thalamic structures in the generation of SWD, with a possible rhythmic triggering by the thalamus.

Using positron-emission-tomography measurements, it has been shown that petit mal absences in humans are associated with a diffuse increase in local cerebral metabolic rates for glucose. Similar results were obtained in GAERS using an autoradiographic method (Nehlig et al., 1991, this volume).

According to their degree of similarity with the human disease, three categories of animal models may be distinguished: (1) isomorphic models with symptoms and occurrence similar to the typical pathology; (2) predictive models showing a similar pattern of response to therapeutic agents, which may lead to the development of new therapies for the disease; and (3) homologous models with a similar aetiology to the human pathology (Kornetsky, 1977).

Similarities with human absence seizures strongly support the epileptic nature of rodent SWD and fully agree with the requirements for an experimental model of petit mal seizures. Rats' spontaneous SWD can actually be considered an isomorphic and predictive model of human generalized non-convulsive (absence) epilepsy, on the basis of EEG, behavioural, and pharmacological data. Neurophysiological and genetic data suggest that these SWD and human absences are likely to be related to similar neural mechanisms and that rats' SWD may therefore be also considered a homologous model of petit mal epilepsy. As the mechanisms involved in the thalamocortical dysfunction in rat and human SWD are still unknown, their analysis in rats may be fruitful in the investigation of the pathogenesis of generalized non-convulsive epilepsy.

References

Aicardi J (1986) Epilepsy in children. Raven Press, New York

Albe-Fessard D, Lombard MC (1983) Use of an animal model to evaluate the origin of and protection against deafferentation pain. In: Bonica JJ (ed) Advances in pain research and therapy, vol 5. Raven Press, New York, pp 691–700

Aldino C, Aporti F, Calderini G, Mazzari S, Zanotti A, Toffano G (1985) Experimental models of aging and quinolinic acid. Meth Find Exp Clin Pharmacol 7: 563–568

Aporti F, Borsato R, Calderini G, Rubini R, Toffano G, Zanotti A, Valzelli L, Goldstein L (1986) Age-dependent spontaneous EEG bursts in rats: effects of brain phosphatidylserine. Neurobiol Aging 7: 115–120

Avoli M (1980) Electroencephalographic and pathophysiologic features of rat parenteral penicillin epilepsy. Exp Neuropharmacol 69: 373–382

Avoli M, Gloor P (1982) Interaction of cortex and thalamus in spike and wave discharges of feline generalized penicillin epilepsy. Exp Neurol 76: 196–217

Avoli M, Gloor P, Kostopoulos G, Gotman J (1983) An analysis of penicillin-induced generalized spike and wave discharges using simultaneous recordings of cortical and thalamus single neurons. J Neurophysiol 50: 819–837

Bassant MH, Cathala F, Court L, Gourmelon P, Hauw JJ (1984) Experimental scrapie in rats: first electrophysiological observations. Electroencephalogr Clin Neurophysiol 57: 541–547

Bassant MH, Court L, Cathala F (1987) Impairment of the cortical and thalamic electrical activity in scrapie-infected rats. Electroencephalogr Clin Neurophysiol 66: 307–316

Berkovic SF, Andermann F, Andermann E, Gloor P (1987) Concepts of absence epilepsies: discrete syndromes or biological continuum? Neurology 37: 993–1000

Bernasconi R, Marescaux C, Vergnes M, Klebs K, Klein M, Martin P, Portet C, Maitre L, Schmutz M (1988) Evaluation of the anticonvulsant and biochemical activity of CGS 8216 and CGS 9896 in animal models. J Neural Transm 71: 11–27

Bernasconi R, Lauber J, Marescaux C, Vergnes M, Martin P, Rubio V, Leonhardt T, Reymann N, Bittiger H (1992) Experimental absence seizures: potential role of gamma-hydroxybutyric acid and $GABA_B$ receptors (this volume)

Billiard M (1982) Epilepsies and the sleep-wake cycle. In: Sterman MB, Shouse MN, Passouant P (eds) Sleep and epilepsy. Academic Press, New York, pp 269–286

Buonamici M, Maj R, Pagani F, Rossi AC, Khazan N (1986) Tremor at rest episodes in unilaterally 6-OHDA-induced substantia nigra lesioned rats: EEG-EMG and behavior. Neuropharmacology 25: 323–325

Buzsáki G, Bickford RG, Armstrong DM, Ponomareff G, Chen KS, Ruiz R, Thal LJ, Gage FH (1988a) Electric activity in the neocortex of freely moving young and aged rats. Neuroscience 26: 735–744

Buzsáki G, Bickford RG, Ponomareff G, Thal LJ, Mandel RJ, Gage FH (1988b) Nucleus basalis and thalamic control of neocortical activity in the freely moving rat. J Neurosci 8: 4007–4026

Buzsáki G, Laszlovszky I, Lajtha A, Vadász C (1990a) Spike-and-wave neocortical patterns in rats: genetic and aminergic control. Neuroscience 38: 323–333

Buzsáki G, Smith A, Berger S, Fisher LJ, Gage FH (1990b) Petit mal epilepsy and parkinsonian tremor: hypothesis of a common pacemaker. Neuroscience 36: 1–14

Chocholová L (1976) Effect of diazepam on the electroencephalographic pattern and vigilance of unanaesthetized and curarized rats with a chronic cobalt-gelatin focus. Physiol Bohemoslov 25: 129–137

Chocholová L (1983) Incidence and development of rhythmic episodic activity in the electroencephalogram of a large rat population under chronic conditions. Physiol Bohemoslov 32: 10–18

Chocholová L, Radil-Weiss T (1970) The level of vigilance and the EEG manifestations of cortical cobalt foci in rats. Physiol Bohemoslov 19: 385–396

Chocholová L, Radil-Weiss T (1973) Effect of diphenylhydantoin on paroxysmal EEG activity induced by cortical cobalt focus. Activ Nerv Super 15: 70–76

Coenen AML, Van Luijtelaar ELJM (1987) The WAG/Rij rat model for absence epilepsy: age and sex factors. Epilepsy Res 1: 297–301

Depaulis A (1983) A microcomputer method for behavioral data acquisition and subsequent analysis. Pharmacol Biochem Behav 19: 729–732

Depaulis A (1992) The inhibitory control of the substantia nigra over generalized non-convulsive seizures in the rat (this volume)

Depaulis A, Bourguignon JJ, Marescaux C, Vergnes M, Schmitt M, Micheletti G, Warter JM (1988a) Effects of gamma-hydroxybutyrate and gamma-butyrolactone derivatives on spontaneous generalized non-convulsive seizures in the rat. Neuropharmacology 27: 683–689

Depaulis A, Vergnes M, Marescaux C, Lannes B, Warter JM (1988b) Evidence that activation of GABA receptors in the subtantia nigra suppresses spontaneous spike-and-wave discharges in the rat. Brain Res 448: 20–29

Depaulis A, Snead OC III, Marescaux C, Vergnes M (1989) Suppressive effects of intranigral injection of muscimol in three models of generalized non-convulsive epilepsy induced by chemical agents. Brain Res 498: 64–72

Depaulis A, Liu Z, Vergnes M, Marescaux C, Micheletti G, Warter JM (1990a) Suppression of spontaneous generalized non-convulsive seizures in the rat by microinjection of GABA antagonists into the superior colliculus. Epilepsy Res 5: 192–198

Depaulis A, Vergnes M, Liu Z, Kempf E, Marescaux C (1990b) Involvement of the nigral output pathways in the inhibitory control of the substantia nigra over generalized non-convulsive seizures in the rat. Neuroscience 39: 339–349

Doose H, Gerken H, Horstmann T, Völzke E (1973) Genetic factors in spike-and-wave absences. Epilepsia 14: 57–75

Dow RS, Fernández-Guardiola A, Manni E (1962) The influence of the cerebellum on experimental epilepsy. Electroencephalogr Clin Neurophysiol 14: 383–398

Dusser AE, Peroutka SJ (1990) Neurotransmitter receptors in adult tottering (tg/tg) mice. Epilepsia 31: 378–381

Fariello RG, Golden GT (1987) The THIP-induced model of bilateral synchronous spike and wave in rodents. Neuropharmacology 26: 161–165

Fariello RG, Golden GT, Black JA (1980) Potentiation of a feline model of corticoreticular epilepsy by systematically administered inhibitory amino acids. In: Canger R, Angeleri F, Penry JK (eds) Advances in Epileptology, XIth Epilepsy International Symposium. Raven Press, New York, pp 339–342

Frey HH, Voits M (1991) Effect of psychotropic agents on a model of absence epilepsy in rats. Neuropharmacology 30: 651–656

Gloor P, Fariello RG (1988) Generalized epilepsy: some of its cellular mechanisms differ from those of focal epilepsy. TINS 11: 63–68

Gloor P, Metrakos J, Metrakos K, Andermann E, Van Gelder N (1982) Neurophysiological, genetic and biochemical nature of the epileptic diathesis. In: Broughton RJ (ed) Henri Gastaut and the Marseille School's contribution to the neurosciences. Elsevier Biomedical Press, Amsterdam, pp 45–56 (EEG [Suppl] 35)

Guey J, Bureau M, Dravet C, Roger J (1969) A study of the rhythm of petit mal absences in children in relation to prevailing situations. The use of EEG telemetry during psychological examinations, school exercises and periods of inactivity. Epilepsia 10: 441–451

Hammond EJ, Villarreal HJ, Wilder BJ (1979) Distinction between normal and epileptic rhythms in rodent sensorimotor cortex. Epilepsia 20: 511–518

Heller AH, Dichter MA, Sidman RL (1983) Anticonvulsant sensitivity of absence seizures in the tottering mutant mouse. Epilepsia 25: 25–34

Jensen LH, Marescaux C, Vergnes M, Micheletti G, Petersen EN (1984) Antiepileptic action of the β-carboline ZK 91296 in a genetic petit mal model in rats. Eur J Pharmacol 102: 521–524

Jung R (1962) Blocking of petit-mal attacks by sensory arousal and inhibition of attacks by an active change in attention during the epileptic aura. Epilepsia 3: 435–437

Kaplan BJ (1985) The epileptic nature of rodent electrocortical polyspiking is still unproven. Exp Neurol 88: 425–436

Kaplan BJ, Seyfried TN, Glaser GH (1979) Spontaneous polyspike discharges in an epileptic mutant mouse (tottering). Exp Neurol 66: 577–586

King GA (1979) Effects of systematically applied GABA agonists and antagonists on wave-spike ECoG activity in rat. Neuropharmacology 18: 47–55

King GA, Burnham WM (1980) Effects of d-amphetamine and apomorphine in a new animal model of petit mal epilepsy. Psychopharmacology 69: 281–285

King GA, Burnham WM (1982) α_2-adrenergic antagonists suppress epileptiform EEG activity in a petit mal seizure model. Life Sci 30: 293–298

Kleinlogel H (1985) Spontaneous EEG paroxysms in the rat: effects of psychotropic and alphaadrenergic agents. Neuropsychobiology 13: 206–213

Klingberg F, Pickenhain L (1968) Das Auftreten von "Spindelentladungen" bei der Ratte in Beziehung zum Verhalten. Acta Biol Med Gem 20: 45–54

Knight AR, Bowery NG (1992) GABA$_B$ receptors in rats with spontaneous generalized non-convulsive epilepsy (this volume)

Kohler M, Klingberg F (1969) Auslösung von Spindelaktivität im Noekortex der Ratte durch Reizung des ventralen Thalamuskernes. Acta Biol Med Gem 23: 99–110

Kornetsky C (1977) Animal models: promises and problems. In: Hanin I, Usdin E (eds) Animal models in psychiatry and neurology. Pergamon Press, Oxford, pp 1–7

Kostopoulos G, Veronikis DK, Efthimiou I (1987) Caffeine blocks absence seizures in the tottering mutant mouse. Epilepsia 28: 415–420

Lannes B, Micheletti G, Vergnes M, Marescaux C, Depaulis A, Warter JM (1988) Relationship between spike-wave discharges and vigilance levels in rats with spontaneous petit mal-like epilepsy. Neurosci Lett 94: 187–191

Lannes B, Vergnes M, Marescaux C, Depaulis A, Micheletti G, Warter JM, Kempf E (1991) Lesions of noradrenergic neurons in rats with spontaneous generalized non-convulsive epilepsy. Epilepsy Res 9: 79–85

Lennox WG, Lennox MA (1960) Epilepsy and related disorders 1. Little Brown and Co, Boston

Libet B, Gleason CA, Wright EW, Feinstein B (1977) Suppression of an epileptiform type of electrocortical activity in the rat by stimulation in the vicinity of locus coeruleus. Epilepsia 18: 451–455

Libouban S, Oswaldo-Cruz E (1958) Quelques observations relatives aux activitiés évoquées et spontanées du cerveau du rat blanc. J Physiol (Paris) 50: 380–383

Liu Z, Seiler N, Marescaux C, Depaulis A, Vergnes M (1990) Potentiation of gamma-vinyl GABA (vigabatrin) effects by glycine. Eur J Pharmacol 182: 109–115

Liu Z, Snead OC III, Vergnes M, Depaulis A, Marescaux C (1991a) Intrathalamic injections of gamma-hydroxybutyric acid increase genetic absence seizures in rats. Neurosci Lett 125: 19–21

Liu Z, Vergnes M, Depaulis A, Marescaux C (1991b) Evidence for a critical role of GABAergic transmission within the thalamus in the genesis and control of absence seizures in the rat. Brain Res 545: 1–7

Loiseau P, Cohadon F (1970) Le petit mal et ses frontières. Masson, Paris

Löscher W, Nau H, Marescaux C, Vergnes M (1984) Comparative evaluation of anticonvulsant and toxic potencies of valproic acid and 2-en-valproic acid in different animal models of epilepsy. Eur J Pharmacol 99: 211–218

Luna D, Dulac O, Pajot N, Beaumont D (1989) Vigabatrin in the tratment of childhood epilepsies: a single-blind placebo-controlled study. Epilepsia 30: 430–437

McLachlan RS, Avoli M, Gloor P (1984a) Transition from spindles to generalized spike and wave discharges in the cat: simultaneous single-cell recordings in cortex and thalamus. Exp Neurol 85: 413–425

McLachlan RS, Gloor P, Avoli M (1984b) Differential participation of some "specific" and "non-specific" thalamic nuclei in generalized spike and wave discharges of feline generalized penicillin epilepsy. Brain Res 307: 277–287

McQueen JK, Woodbury DM (1975) Attempts to produce spike-and-wave complexes in the electro-corticogram of the rat. Epilepsia 16: 295–299

Maître M, Hechler V, Vayer P, Gobaille S, Cash CD, Schmitt M, Bourguignon JJ (1990) A specific gamma-hydroxybutyrate receptor ligand possesses both antagonistic and anticonvulsant properties. J Pharmacol Exp Ther 255: 657–663

Marescaux C, Micheletti G, Vergnes M, Depaulis A, Rumbach L, Warter JM (1984a) A model of chronic spontaneous petit mal-like seizures in the rat: comparison with pentylenetetrazol-induced seizures. Epilepsia 25: 326–331

Marescaux C, Micheletti G, Vergnes M, Depaulis A, Rumbach L, Warter JM (1984b) Biphasic effects of Ro 15-1788 on spontaneous petit mal-like seizures in rats. Eur J Pharmacol 102: 355–359

Marescaux C, Vergnes M, Micheletti G (1984c) Antiepileptic drug evaluation in a new animal model: spontaneous petit mal epilepsy in the rat. Fed Proc 43: 280–281

Marescaux C, Vergnes M, Micheletti G, Depaulis A, Reis J, Rumbach L, Warter JM, Kurtz D (1984d) Une forme génétique d'absences petit mal chez le rat Wistar. Rev Neurol (Paris) 140: 63–66

Marescaux C, Micheletti G, Vergnes M, Rumbach L, Warter JM (1985) Diazepam antagonizes GABAmimetics in rats with spontaneous petit mal-like epilepsy. Eur J Pharmacol 113: 19–24

Marescaux C, Vergnes M, Jensen LH, Petersen E, Depaulis A, Micheletti G, Warter JM (1987a) Bidirectional effects of beta-carbolines in rats with spontaneous petit mal-like seizures. Brain Res Bull 19: 327–335

Marescaux C, Vergnes M, Micheletti G, Rumbach L, Warter JM (1987b) Electro-corticographic and behavioral effects of repeated injections of non-convulsant doses of pentylenetetrazol in the rat. In: Wolf P, Dam M, Janz D, Dreifuss FE (eds) Advances in epileptology, vol 16. Raven Press, New York, pp 67–69

Marescaux C, Vergnes M, Depaulis A, Micheletti G, Warter JM (1992a) Neurotransmission in rats' spontaneous generalized nonconvulsive epilepsy. In: Avanzini G, et al (eds) Neurotransmitters in epilepsy. Epilepsy Res [Suppl] (in press)

Marescaux C, Vergnes M, Bernasconi R (1992b) $GABA_B$ receptor antagonists: potential new anti-absence drugs (this volume)

Meldrum B, Horton R (1980) Effects of the bicyclic GABA agonist, THIP, on myoclonic and seizure responses in mice and baboons with reflex epilepsy. Eur J Pharmacol 61: 231–237

Metrakos K, Metrakos JD (1974) Genetics of epilepsy. In: Vinken PJ, Bruyn GW (eds) Handbook of clinical neurology, vol 15. North-Holland, Amsterdam, pp 429–439

Micheletti G, Marescaux C, Vergnes M, Rumbach L, Warter JM (1985a) Effects of GABAmimetics and GABA antagonists on spontaneous nonconvulsive seizures in Wistar rats. In: Bartholini G, Bossi L, Lloyd KG, Morselli ML (eds) L.E.R.S. Monograph series, vol 3. Raven Press, New York, pp 129–137

Micheletti G, Vergnes M, Marescaux C, Reis J, Depaulis A, Rumbach L, Warter JM (1985b) Antiepileptic drug evaluation in a new animal model: spontaneous petit mal epilepsy in the rat. Arzneimittelforschung/Drug Res 35: 483–485

Micheletti G, Warter JM, Marescaux C, Depaulis A, Tranchant C, Rumbach L, Vergnes M (1987) Effects of drugs affecting noradrenergic neuro-transmission in rats with spontaneous petit mal-like seizures. Eur J Pharmacol 135: 397–402

Mirsky AF, Duncan CC, Myslobodsky MS (1986) Petit mal epilepsy: a review and integration of recent information. J Clin Neurophysiol 3: 179–208

Mouritzen Dam A, Marescaux C, Vergnes M, Dam M (1989) Thalamus and epilepsy. In: Manelis J, Bental E, Loeber JN, Dreifuss FE (eds) Advances in epileptology, vol 17. Raven Press, New York, pp 63–66

Musgrave J, Gloor P (1980) The role of the corpus callosum in bilateral interhemispheric synchrony of spike and wave discharge in feline generalized penicillin epilepsy. Epilepsia 21: 369–378

Nehlig A, Vergnes M, Marescaux C, Boyet S, Lannes B (1991) Local cerebral glucose utilization in rats with petit mal-like seizures. Ann Neurol 29: 72–77

Nehlig A, Vergnes M, Marescaux C, Boyet S (1992) Mapping of cerebral energy metabolism in rats with genetic generalized non convulsive epilepsy (this volume)

Noebels JL (1984) A single gene error of noradrenergic axon growth synchronizes central neurones. Nature 310: 409–411

Noebels JL, Sidman RL (1979) Inherited epilepsy: spike-wave and focal motor seizures in the mutant mouse tottering. Science 204: 1334–1336

Noebels JL, Qiao X, Bronson RT, Spencer C, Davisson MT (1990) Stargazer: a new neurological mutant on chromosome 15 in the mouse with prolonged cortical seizures. Epilepsy Res 7: 129–135

Patel S, Chapman AG, Millan MH, Meldrum BS (1988) Epilepsy and excitatory amino acids antagonists. In: Lodge D (ed) Excitatory amino acids in health and disease. Wiley, Chichester, pp 353–378

Peeters BWMM, Sporen WPJM, van Luijtelaar ELJM, Coenen AML (1988) The WAG/Rij rat model for absence epilepsy: anticonvulsant drug evaluation. Neurosci Res Commun 2: 93–97

Peeters BWMM, van Rijn CM, van Luijtelaar ELJM, Coenen AML (1989) Antiepileptic and behavioural actions of MK-801 in an animal model of spontaneous absence epilepsy. Epilepsy Res 3: 178–181

Peeters BWMM, Kerbusch JML, van Luijtelaar ELJM, Vossen JMH, Coenen AML (1990a) Genetics of absence epilepsy in rats. Behav Genet 20: 453–460

Peeters BWMM, van Rijn CM, Vossen JMH, Coenen AML (1990b) Involvement of NMDA receptors in non-convulsive epilepsy in WAG/Rij rats. Life Sci 47: 523–529

Prince D, Farrell D (1969) "Centrencephalic" spike-wave discharges following parenteral penicillin injection in the cat. Neurology 19: 309–310

Qiao X, Noebels JL (1991) Genetic and phenotypic heterogeneity of inherited spike-wave epilepsy: two mutant gene loci with independent cerebral excitability defects. Brain Res 555: 43–50

Quesney LF, Reader T (1984) Role of cortical catecholamine depletion in the genesis of epileptic photosensitivity. In: Fariello RG, Morselli PL, Lloyd KG, Quesney LF, Engel J (eds) Neurotransmitters, seizures and epilepsy II. Raven Press, New York, pp 11–21

Radil T, Chocholová L, Roldán E (1982) The influence of sleep-waking states on EEG manifestations of experimental epileptoid foci. In: Sterman MB, Shouse MN, Passouant P (eds) Sleep and epilepsy. Academic Press, New York, pp 73–88

Robinson PF, Gilmore SA (1980) Spontaneous generalized spike-wave discharges in the electrocorticograms of albino rats. Brain Res 201: 452–458

Roldán E, Radil-Weiss T, Chocholová L (1970) Paroxysmal activity of hippocampal and thalamic epileptogenic foci and induced or spontaneous changes of vigilance. Exp Neurol 29: 121–130

Ryan LJ (1984) Characterization of cortical spindles in DBA/2 and C5BL/6 inbred mice. Brain Res Bull 13: 549–558

Ryan LJ (1985a) Cholinergic regulation of neocortical spinding in DBA/2 mice. Exp Neurol 89: 372–381

Ryan LJ (1985b) Catecholamine regulation of neocortical spindling in DBA/2 mice. Behav Brain Res 16: 103–115

Ryan LJ, Sharpless SK (1979) Genetically determined spontaneous and pentylenetetrazol-induced brief spindle episodes in mice. Exp Neurol 66: 493–508

Sasa M, Ohno Y, Ujihara H, Fujita Y, Yoshimura M, Takaori S, Serikawa T, Yamada J (1988) Effects of antiepileptic drugs on absence-like and tonic seizures in the spontaneous epileptic rat, a double mutant rat. Epilepsia 29: 505–513

Schickerová R, Mareš P, Trojan S (1984) Correlation between electrocorticographic and motor phenomena induced by pentamethylenetetrazol during ontogenesis in rats. Exp Neurol 84: 153–164

Semba K, Komisaruk BR (1984) Neural substates of two different rhythmical vibrissal movements in the rat. Neuroscience 12: 761–774

Semba K, Szechtamn H, Komisaruk BR (1980) Synchrony among rhythmical facial tremor neocortical "alpha" waves and thalamic non-sensory neuronal bursts in intact awake rats. Brain Res 195: 281–298

Serikawa T, Ohno Y, Sasa M, Yamada J, Takaori S (1987) A new model of petit mal epilepsy: spontaneous spike and wave discharges in tremor rats. Lab Anim 21: 68–71

Smith KA, Bierkamper GG (1990) Paradoxical role of GABA in a chronic model of petit mal (absence)-like epilepsy in the rat. Eur J Pharmacol 176: 45–55

Snead OC III (1978) Gamma hydroxybutyrate in the monkey. Neurology 28: 636–642

Snead OC III (1988) Gamma-hydroxybutyrate model of generalized absence seizures: Further characterization and comparison with other absence models. Epilepsia 29: 361–368

Snead OC III (1992) Pharmacological models of generalized absence seizures in rodents (this volume)

Snead OC III, Hechler V, Vergnes M, Marescaux C, Maitre M (1990) Increased gamma-hydroxybutyric acid receptors in thalamus of a genetic animal model of petit mal epilepsy. Epilepsy Res 7: 121–128

Stefan H, Plouin P, Fichsel H, Jalin C, Burr W (1988) Progabide for previously untreated absence epilepsy. Epilepsy Res 2: 132–136

Steriade M, Deschenes M (1984) The thalamus as a neuronal oscillator. Brain Res Rev 8: 1–63

Timo-Iaria C, Negráo N, Schmidek WR, Hoshino K, Lobato de Menezes CE, Leme da Rocha T (1970) Phases and stated of sleep in the rat. Physiol Behav 5: 1057–1062

Van Luijtelaar ELJM, Coenen AML (1986) Two types of electrocortical paroxysms in an inbred strain of rats. Neurosci Lett 70: 393–397

Vanderwolf CH (1975) Neocortical and hippocampal activation in relation to behavior: effects of atropine, eserine, phenothiazines, and amphetamine. J Comput Physiol Psychol 88: 300–323

Vanderwolf CH, Robinson TE (1981) Reticulo-cortical activity and behavior: a critique of the arousal theory and a new synthesis. Behav Brain Sci 4: 459–514

Vergnes M, Marescaux C (1992) Cortical and thalamic lesions in rats with genetic absence epilepsy (this volume)

Vergnes M, Marescaux C, Micheletti G, Reis J, Depaulis A, Rumbach L, Warter JM (1982) Spontaneous paroxysmal electroclinical patterns in rat: a model of generalized non-convulsive epilepsy. Neurosci Lett 33: 97–101

Vergnes M, Marescaux C, Micheletti G, Depaulis A, Rumbach L, Warter JM (1984) Enhancement of spike and wave discharges by GABAminetic drugs in rats with spontaneous petit mal-like epilepsy. Neurosci Lett 44: 91–94

Vergnes M, Marescaux C, Micheletti G, Rumbach L, Warter JM (1985) Blockage of "antiabsence" activity of sodium valproate by THIP in rats with petit mal-like seizures. J Neural Transm 63: 133–141

Vergnes M, Marescaux C, Depaulis A, Micheletti G, Warter JM (1986) Ontogeny of spontaneous petit mal-like seizures in Wistar rats. Dev Brain Res 30: 85–87

Vergnes M, Marescaux C, Depaulis A, Micheletti G, Warter JM (1987) Spontaneous spike and wave discharges in thalamus and cortex in a rat model of genetic petit mal-like seizures. Exp Neurol 96: 127–136

Vergnes M, Marescaux C, Lannes B, Depaulis A, Micheletti G, Warter JM (1989) Interhemispheric desynchronization of spontaneous spike-wave discharges by corpus callosum transection in rats with petit mal-like epilepsy. Epilepsy Res 4: 8–13

Vergnes M, Marescaux C, Depaulis A (1990a) Mapping of spontaneous spike and wave discharges in Wistar rats with genetic generalized non-convulsive epilepsy. Brain Res 523: 87–91

Vergnes M, Marescaux C, Depaulis A, Micheletti G, Warter JM (1990b) Spontaneous spike-and-wave discharges in Wistar rats: a model of genetic generalized nonconvulsive epilepsy. In: Avoli M, Gloor P, Kostopoulos G, Naquet R (eds) Generalized epilepsy. Birkäuser, Boston, pp 238–253

Vergnes M, Marescaux C, Boehrer A, Depaulis A (1991) Are rats with genetic absence epilepsy behaviorally impaired? Epilepsy Res 9: 97–104

Wahle H, Frey HH (1990) Development of tolerance to the anticonvulsant effect of valproate but not to ethosuximide in a rat model of absence epilepsy. Eur J Pharmacol 181: 1–8

Warter JM, Vergnes M, Depaulis A, Tranchant C, Rumbach L, Micheletti G, Marescaux C (1988) Effects of drugs affecting dopaminergic neurotransmission in rats with spontaneous petit mal-like seizures. Neuropharmacology 27: 269–274

Warter JM, Tranchant C, Marescaux C, Depaulis A, Lannes B, Vergnes M (1990) Immediate effects of 14 non MAOI antidepressants in rats with spontaneous petit mal-like seizures. Prog NeuroPsychopharmacol Biol Psychiatry 14: 261–270

Whishaw IQ, Vanderwolf CH (1971) Hippocampal EEG and behavior: effects of variation in body temperature and relation of EEG to vibrissae movement, swimming and shivering. Physiol Behav 6: 391–397

Williams D (1953) A study of thalamic and cortical rhythms in petit mal. Brain 76: 50–69

Willmore LJ, Sypert GW, Munson JB, Hurd RW (1978) Chronic focal epileptiform discharges induced by injection of iron into rat and cat cortex. Science 200: 1501–1502

Authors' address: Dr. C. Marescaux, Service de Neurologie I, Hôpital Universitaire, 1 place de l'Hôpital, F-67091 Strasbourg Cedex, France.

J Neural Transm (1992) [Suppl] 35: 71–83
© Springer-Verlag 1992

Cortical and thalamic lesions in rats with genetic absence epilepsy

M. Vergnes and **C. Marescaux**

Laboratoire de Neurophysiologie et Biologie des Comportements,
Centre de Neurochimie du CNRS, Strasbourg, France

Summary. In generalized, non-convulsive, absence epilepsy, spike-and-wave discharges (SWD) are recorded in both the cortex and the thalamus. The effect of various cortical and thalamic lesions on the occurrence of spontaneous SWD was examined in rats from a strain with genetic absence epilepsy.

Cortical ablations suppressed SWD recorded in the thalamus. KCl induced unilateral cortical spreading depression and transiently suppressed SWD in the ipsilateral cortex and thalamus; SWD recovered simultaneously in both structures.

Bilateral thalamic lesions of the anterior nuclei, the ventromedial nuclei, the posterior area, or lesion of the midline nuclei did not suppress cortical SWD. However, large lesions of the lateral thalamus, including the specific relay and reticular nuclei, definitely suppressed ipsilateral SWD, and pentylenetetrazol, THIP or gammabutyrolactone failed to restore the cortical SWD.

These results demonstrate that the neocortex and the specific thalamic nuclei are both necessarily involved in the generation of SWD in absence epilepsy.

Introduction

Epileptic seizures typical of generalized, non-convulsive epilepsy occur spontaneously in a strain of Wistar rats selected in our laboratory and called "genetic absence epilepsy rats from Strasbourg" (GAERS). Behavioral, electroencephalographic (EEG), pharmacological and genetic characteristics are similar to those of the human disease, also termed petit mal or absence epilepsy (Vergnes et al., 1982, 1990). During the seizures, which occur spontaneously in waking animals, bilaterally synchronous spike and wave discharges (7–10 c/s, mean duration 20 s/min) are recorded from the EEG of these animals. In addition, localized EEG recordings from various brain regions have shown a predominance of the SWD in the neocortex and thalamus. The SWD were particularly large in the frontoparietal cortex and the relay nuclei of the lateral thalamus (Vergnes et al., 1987), suggesting that these structures play a predominant role in the development of SWD. In humans, SWD were similarly recorded over the cortex and in the thalamus (Williams, 1953). These data are in agreement with electrophysiological results observed in generalized epilepsy induced by penicillin in the cat (Avoli and Gloor, 1982; Avoli et al., 1983; Mc Lachlan et al., 1984). In this feline model, cortical and thalamic structures were shown to be intimately associated in the generation of SWD.

In the present experiments various cortical and thalamic areas were lesioned in GAERS to ascertain which structures are essential to the occurrence of spontaneous SWD and to characterize cortico-thalamic relationships in the generation of SWD. Since the reciprocal thalamo-cortical connections are ipsilateral, the effect of a cortical or a thalamic lesion is reflected in the EEG of the ipsilateral thalamus or cortex respectively. In order to prevent the spread of the SWD from one hemisphere to the other (Vergnes et al., 1989), the corpus callosum was transected in animals with unilateral lesions, the unlesioned side serving as a control.

Cortical lesions

The cortex was lesioned in two different ways. In the first series of experiments, cortical ablation was performed by suction. With this technique the lateral parts of the cortex cannot be removed completely without also damaging neighboring structures.

In the second group of experiments a functional and acute elimination of the cortex was effected, by local application of a KCl solution, which induces a spreading depression that extends transiently all over the cortex (Bures et al., 1974).

Cortical ablation

Methods

Six male animals of the GAERS strain were anesthetized with pentobarbital (40 mg/kg ip) and placed in a stereotaxic apparatus. The bone overlying the cortex was removed with a dental drill, preserving the midline skull and a small strip of bone through which the ipsilateral thalamic electrode had to be lowered. The dorsal part of the cortex was extensively removed over the right hemisphere, by suction through a pipette connected to a water pump.

A bipolar electrode made of twisted enameled stainless steel, with 1 mm dorsoventral distance between the two poles, was then placed stereotoxically into the lateral thalamus at the following coordinates in mm, with reference to lambda: AP = 5, ML = 2, DV = 6,5. On the opposite side, two single

contact electrodes were screwed into the skull over the fronto-parietal cortex.

All electrodes were connected to a microconnector and embedded in acrylic cement. The animals were treated with a single injection of extencillin to prevent infection. The cortical and thalamic EEG was recorded regularly during the two postoperative months. At the end of that time, the animals were killed with an overdose of pentobarbital. The brains were removed and serial brain slices were prepared for histological control of lesion and electrode location.

Results

In all animals SWD were absent in the right thalamus following the ipsilateral decortication. On the opposite non lesioned side, the SWD appeared normal. In 4 animals, the SWD reappeared in the right thalamus between the 15th and 30th day after cortical lesioning. The SWD usually occurred separately in the right and left hemisphere, this interhemispheric desynchronization resulting from an associated damage to the corpus callosum.

In two animals no SWD were recorded in the right thalamus during the 2 months of survival, and the baseline EEG was apparently normal with fast low voltage activity.

Histology

The recording electrodes were well localized in the ventrolateral thalamus. The frontal cortex was almost completely removed in all animals. The parietal cortex was removed dorsally but was spared laterally to various extents. The lesion also involved the dorsal occipital cortex. The suppression of ipsilateral thalamic SWD was related to the extent of the cortical lesion, especially of the lateral parietal cortex, the rats which recovered SWD having the smallest lesions. The corpus callosum was damaged in all animals. In addition, damaged neurons were found in lateral relay nuclei at a distance from the cortical lesion. These lesions probably resulted from retrograde degeneration of thalamic neurons projecting to the lesioned cortex.

Cortical spreading depression

Methods

Two male rats of the GAERS strain were used. First, the corpus callosum and medial thalamus were transected with a small surgical blade mounted on a stereotaxic holder. After removal of the midline skull with a dental drill, the blade was lowered at AP = 2, DV = 4 (coordinates in mm with the lambda as reference) into the midline of the brain and then moved anteriorly to AP = 4.5, where it was lowered to DV = 7 and moved to AP = 7. The blade was then raised to DV = 4.5, moved to AP = 12, and removed dorsally. After one week's recovery the rats were implanted bilaterally with a bipolar electrode in the lateral thalamus (AP = 6, ML = 2.5, DV = 6.5), two single contact electrodes over the cortex (AP = 10.5 and 4, ML = 3) and a stainless-steel guide cannula (AP = 8, ML = 4, DV = 1; OD = 0.4, ID = 0,3), which was attached to the connector and was also used as a cortical electrode. Stylets of the same length were left in place in the cannula and were replaced for injections by an inner cannula with a smaller diameter (OD = 0.28, ID = 0.18), but the same length as the guide cannula. Doses of 1 or 0.5 μl 25% KCl solution, or 0.9% ClNa as control, were injected unilaterally at a rate of 1 μl/min with a microsyringe (Hamilton, 1 μl) through polyethylene tubing fixed to the injection cannula, which was left in place for 30 s after the completion of the injection. The rat was gently immobilized during the injection, then immediately connected to the recording apparatus (Alvar), and the EEG was recorded.

In further experiments the animals were pretreated with drugs which potentiate SWD (Vergnes et al., 1984; Warter et al., 1988; Depaulis et al., 1988). THIP, a GABA A agonist (7.5 mg/kg), haloperidol, a dopamine antagonist (1 mg/kg) or gamma-butyrolactone, an agonist of gamma-hydroxybutyrate receptors (GBL, 250 mg/kg) was injected ip. When the SWD became permanent, 1 μl of KCl solution was applied unilaterally on the cortex.

After 6 months' survival, the animals were killed and the brains processed for histological examination.

Results

The SWD recorded from the left and right cortex and thalamus were completely dissociated as a consequence of the medial transection. They occurred independently and were totally asynchronous in both hemispheres. In contrast, the SWD were synchronized and simultaneous in the cortex and thalamus of the same hemisphere (Fig. 1).

Each animal received four unilateral cortical injections of 0.5 or 1 μl of KCl solution through each cannula. KCl application provoked an immediate flattening of the ipsilateral cortical EEG, whereas the thalamic baseline EEG appeared normal. A total suppression of the SWD was obtained on that side in both cortex and thalamus (Fig. 1). The EEG and SWD on the contralateral side were unchanged. The cortical EEG on the injected side recovered first with slow waves and then progressively with appearance of increasing frequencies. The EEG appeared normal after a period of 10 to 20 min. However, SWD reappeared only later, after delays varying from 37

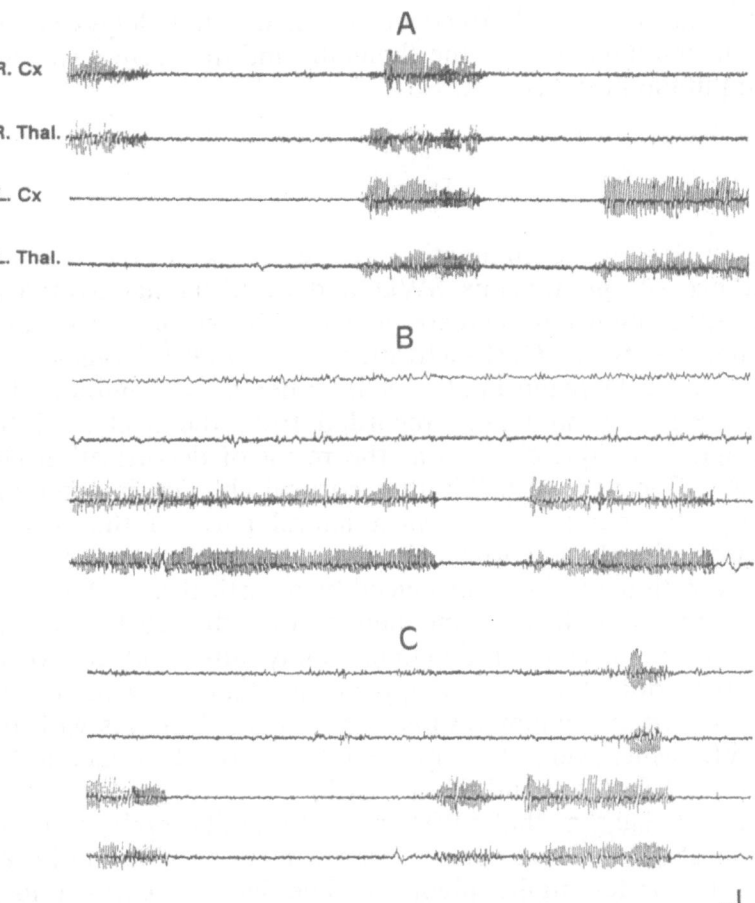

R. Cx

R. Thal.

L. Cx

L. Thal.

Fig. 1. EEG recorded simultaneously in the right cortex (R. Cx), right thalamus (R. Thal), left cortex (L. Cx) and left thalamus (L. Thal) in a rat with a transection of the corpus callosum and the midline thalamus. **A** Control EEG showing the dissociation of the SWD in the left and right hemisphere. **B** 5 min after application of 1 μl of KCl solution on the right cortex producing a spreading depression over the right cortex. SWD are suppressed in both the right thalamus and cortex. In the left thalamus and cortex SWD are unchanged. **C** 40 min after the spreading depression the first SWD reappears simultaneously on the right cortex and thalamus. Calibration 1 s., 200 μV

to 40 min after injection of 0.5 μl and from 1 to 2 h after injection of 1 μl. Recovery of SWD always occurred simultaneously in the ipsilateral cortex and thalamus, starting sometimes with a reduced amplitude in the cortex. No thalamic SWD were observed without a concomitant synchronous oscillation in the cortex (Fig. 1). Control injections of saline suppressed ipsilateral SWD for 3 to 7 min. without altering the baseline EEG.

After administration of THIP, haloperidol or GBL, the SWD were much increased in both hemispheres, simultaneously in the cortex and in the thalamus. Cortical application of 1 μl KCl solution immediately suppressed the ipsilateral SWD, as previously described, the contralateral discharges being unchanged. Recovery occurred after delays varying from 30 min to 2 h simultaneously in the thalamus and the cortex, starting with a reduced amplitude in the cortex.

Discussion

Functional or destructive elimination of the cortex in GAERS suppressed the occurrence of spontaneous SWD and of SWD aggravated by prior injection of drugs known to increase petit-mal-like seizures, such as GABA-mimetics, neuroleptics or GHB-activating compounds (Vergnes et al., 1984; Water et al., 1988; Depaulis et al., 1988). When a large amount of cortex is destroyed, SWD are no longer recorded from the ipsilateral thalamus, where retrograde lesions develop as the result of decortication (Ross and Ebner, 1990). Recovery of SWD after cortical ablation in some animals is probably due to sparing of the most lateral parts of the cortex and/or reorganization of thalamo-cortical circuits.

When a functional lesion is produced by a cortical spreading depression, which is supposed to invade one hemicortex, the SWD are transiently suppressed in the ipsilateral thalamus. Only after full recovery of the cortical activity does the SWD reappear simultaneously in the ipsilateral thalamus and cortex. Similar results were obtained in cats with penicillin-induced SWD, which were also suppressed by cortical spreading depression (Gloor et al., 1979; Avoli and Gloor, 1982).

These results suggest that SWD in petit-mal-like epilepsy requires the participation of a functional cortex. However, no specific site in the cortex appears to be preferentially involved. The lack of cortical localization was previously shown in animals with multiple cortical implantation: the SWD were recorded all over the cortex. Moreover, the SWD started alternately from different locations, which varied from one seizure to another, with some predominance of the lateral frontoparietal cortex (Vergnes, unpublished results). However, localized lesions within these areas did not definitively suppress SWD, suggesting that no specific cortical area is critically involved in the development of SWD.

Thalamic lesions

Methods

Electrolytic lesions (2 mA for 20 sec/site at the cathode) were made at various sites in the thalamus. The stereotaxic coordinates are given in mm, with lambda as reference.

Large unilateral or bilateral lesions of the lateral thalamus: AP = 4, 5, 6; ML = 1.5, 2.5, 3.5; DV = 6 − 7. In 4 animals the corpus callosum was also transected (AP 2 − 12; DV 4 − 4.5). Bilateral lesions of the anterior thalamus: AP = 5.5; ML = 1, 2, 3; DV = 6 − 7. Bilateral lesions of the ventromedial thalamus: AP = 4.5, 5.5; ML = 1.3; DV = 7.3. Lesions of the medial thalamus: AP = 4, 5, 6, 7; ML = 0; DV = 6 − 7.

All animals were implanted with 4 single contact electrodes over the fronto-parietal cortex, either immediately or 2 days after the lesion. The cortical left and right EEG was recorded between two ipsilateral electrodes. The general behavior and the weight of the animals were noted.

At the end of the experiment the animals were killed with an overdose of pentobarbital. Serial paraffin sections of the brains were stained with cresyl violet for histological control of the lesion.

Results

Lateral thalamic lesions

In four animals large bilateral lesions of the lateral thalamus were obtained. All animals lost weight as a result of aphagia after the lesion. One died on the 16th and one on the 23th postoperative day. The other two spontaneously recovered feeding and were killed 34 and 56 days after the lesion, when their EEG had become stable. No SWD was ever recorded from the cortex during the survival period. The cortical EEG was first altered, with large slow waves, the voltage of which was reduced over time; however fast activities remained rare.

In four animals with bilateral lesions extending only to the most posterior part of the lateral thalamus, the SWD were transiently altered.

The lateral thalamus was lesioned unilaterally in six rats, four of them with a prior transection of the corpus callosum. In this latter preparation, the unlesioned hemisphere is considered as control, whereas on the lesioned side the effects of the thalamic lesion can be observed. Moreover, the feeding behavior was not affected in these animals, for 46 days after the lesion. No SWD was recorded from the cortex ipsilateral to the thalamic lesion, which had a slow baseline EEG (Fig. 2D). On the unlesioned side the cortical EEG was normal with many SWD. In the two rats without callosal transection very small SWD on the lesioned side were synchronous with the contralateral ones.

Drugs capable of inducing SWD in non-epileptic rats (Marescaux et al., 1984, 1990) were injected ip. in these animals: THIP (10 mg/kg), GBL (170 mg/kg) and pentylenetetrazole (PTZ, 20 mg/kg). These drugs consistently produced a significant increase of SWD on the unlesioned side. However, no SWD ever occurred in the cortex ipsilateral to the thalamic lesion (Fig. 3).

On the other hand, the injection of a higher dose of PTZ (40 mg/kg) provoked clonic movements of the limbs and the body with bilateral paroxysmal discharges on the cortex in unilaterally thalamic-lesioned animals.

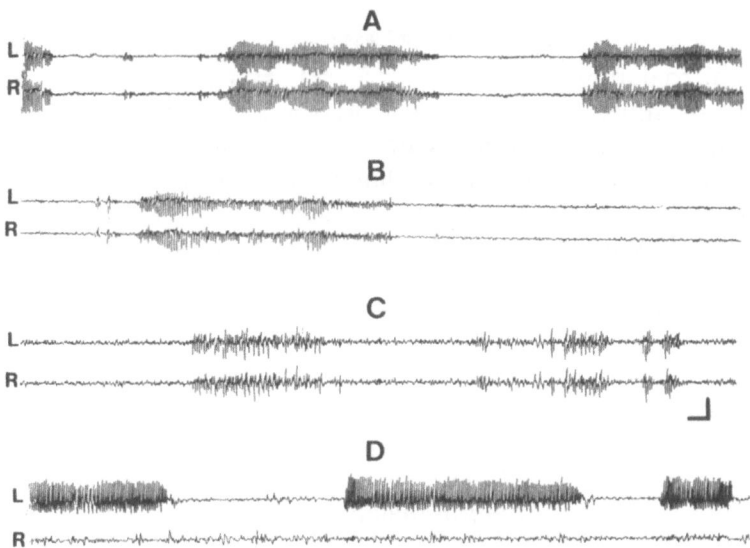

Fig. 2. Left (L) and right (R) cortical EEG after various thalamic lesions. **A** Lesion of the medial thalamic nuclei **B** Bilateral lesion of the ventromedial nucleus **C** Bilateral lesion of the anterior nuclei **D** Unilateral lesion of the right lateral thalamus after transection of the corpus callosum

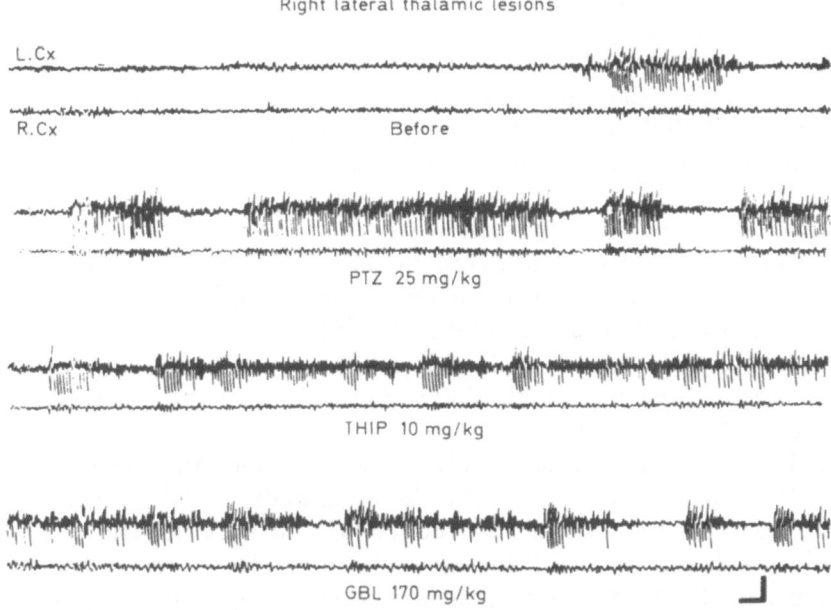

Fig. 3. Left (L. Cx) and right (R. Cx) cortical EEG in a rat with callosal transection and a large lesion of the right lateral thalamus. No SWD ever appeared on the right cortex; **A** before any injection; **B** after PTZ 25 mg/kg; **C** after THIP 10 mg/kg; **D** after GBL 170 mg/kg. PTZ, THIP and GBL increased SWD duration on the unlesioned side, but never induced SWD on the lesioned side

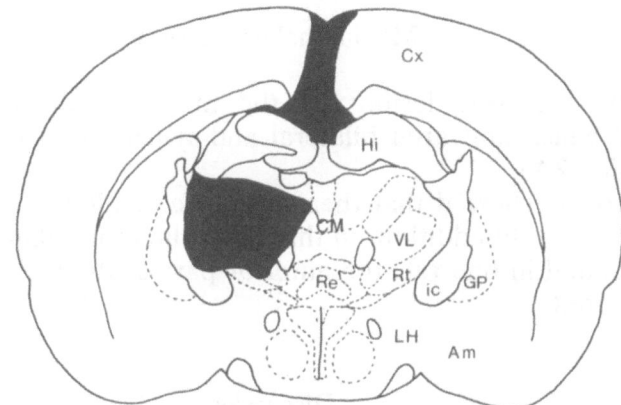

Fig. 4. Coronal section showing a unilateral lesion of the lateral thalamus, with a transection of the corpus callosum, which suppressed the SWD on the ipsilateral cortex. *Am* Amygdala; *CM* central medial n; *Cx* Cortex; *GP* globus pallidus; *Hi* Hippocampus; *ic* internal capsule; *LH* lateral hypothalamus; *Re* reuniens n; *Rt* reticular n; *VL* ventrolateral n. (according to the Atlas of Paxinos and Watson, 1982)

Histological examination of the brains showed extensive lesions of all the lateral part of the anterior to posterior thalamus including the relay nuclei, with partial lesions to the globus pallidus and medial nuclei, in all animals in which SWD were suppressed (Fig. 4). In the cortical projection areas of these thalamic nuclei a suppression of pyramidal cells was apparent. In four animals with lesions restricted to the posterior part of the lateral thalamus, partial damage of the ventroposterolateral nucleus and of the reticular nucleus was observed.

Anterior thalamic lesions

In two rats, bilateral lesions of the anterior thalamus only transiently suppressed the SWD, which reappeared two days after lesioning and then occurred normally (Fig. 2C).

Histological examination of the brains revealed that the anterior thalamic nuclei were bilaterally destroyed, as well as the fornix and the mamillothalamic tract. The anterior part of the ventrolateral and reticular nucleus was affected.

Ventromedial thalamic lesions

In seven rats bilateral lesions of the nucleus ventromedialis and the surrounding area in no way altered the SWD or the baseline EEG (Fig. 2B).

The lesions partially or totally affected the nucleus ventromedialis as well as parts of the bordering zona incerta. The mamillothalamic tract was interrupted bilaterally in four rats.

Medial thalamic lesions

· In four rats extensive lesions of the medial thalamic nuclei did not affect SWD which appeared bilateral and synchronous only one day after lesioning (Fig. 2A).

The histology showed that these lesions extended to all thalamic nuclei along the midline. In addition, in three rats the lesion damaged the anterior commissure, and in one rat, the anterior part of the mesencephalic central gray was altered.

Discussion

The cortical SWD were definitely suppressed after extensive lesioning of the ipsilateral thalamus, including the relay nuclei and the reticular nucleus of the thalamus. The altered cortical EEG with continuous large slow waves gradually reverted to normal, but no SWD ever occurred.

Moreover, drugs such as THIP, GBL or PTZ, which usually induce SWD in non-epileptic rats and potentiate SWD in the epileptic strain, (Marescaux et al., 1984, 1990) never produced SWD on the lesioned side, whereas the SWD on the unlesioned side were markedly increased. These results clearly show that SWD cannot develop from a cortex deprived of its thalamic afferents. Moreover, cell loss, especially in pyramidal layers appears after thalamic deafferentation, which is likely to induce long-term dysfunction of the cortex. However, the clonic seizure induced by a higher dose of PTZ is expressed normally on the deafferented cortex, confirming that different substrates are involved in the various types of seizures induced by increasing doses of PTZ.

None of the anterior nuclei of the thalamus appeared necessary to the occurrence of SWD. These nuclei have distinct connections from other thalamic areas: they do not receive afferents from the thalamic reticular nucleus, whereas they are innervated by a heavy projection from the mamillary body through the mamillothalamic tract (Jones, 1985). The mamillothalamic system has been involved in the propagation of convulsive seizures induced by PTZ (Mirski and Ferendelli, 1986, 1987). But this substrate appears unnecessary for the occurrence and propagation of the SWD, which can be recorded from cortex in rats with lesions of all the anterior thalamic nuclei and/or bilateral interruption of the mamillothalamic tract. Moreover, the SWD are not recorded in the anterior thalamic nuclei (Vergnes et al., 1990).

Similarly, the ventromedial nuclei and the surrounding areas are not necessary for the generation of SWD, which occur with normal frequency and amplitude after lesions of the ventromedial thalamus. The ventromedial nucleus receives a GABAergic projection from the substantia nigra (Di Chiara et al., 1979). Our results suggest that this pathway is not involved in the nigral control of SWD (Depaulis et al., 1990).

The midline nuclei of the thalamus were necessary neither for the occurrence, nor for the bilateral synchronization of the SWD. No SWD are recorded in these nuclei during absence seizures (Vergnes et al., 1987). However, SWD were elicited by electrical stimulation of the intralaminar and medial nuclei in the cat (Hunter and Jasper, 1949). These results suggested that these nuclei might, in some way, be involved in the genesis of bilateral synchronous SWD. However, it is unlikely that the spontaneous SWD in the rat are triggered from these structures.

Thalamic lesions in the model of generalized penicillin epilepsy in the cat produced effects very similar to our own results in GAERS: only large lesions of the lateralis posterior nuclear group abolished the SWD, whereas lesions of the anterior nuclei, the massa intermedia or the ventromedial thalamus did not suppress penicillin-induced seizures (Pellegrini and Gloor, 1979).

Altogether, these results confirm that the cortex and the thalamus are both intimately involved in the genesis of SWD in petit-mal epilepsy. More precisely, it appears that the thalamic relay nuclei are necessary for SWD to occur. These nuclei are characterized by their reciprocal connectivity with the cortex: corticothalamic connections return from every cortical area to the thalamic nuclei, providing input to that area (Jones, 1985). This organization in a closed loop may furnish the substrate allowing an oscillatory activity, possibly originating in the thalamus, to be amplified and expressed as spike and waves.

En route, the cortico-thalamic fibers give collaterals to the reticular nucleus, which, in turn projects GABAergic efferents on most of the thalamic neurons, thus modulating their activity and possibly controlling their ability to discharge with rhythmic bursts (Steriade and Deschenes, 1984). The function of the thalamic reticular nucleus in the control of SWD has to be further investigated.

Whether the generation of SWD is the result of an excessive cortical excitability, as was proposed in regard to feline generalized penicillin epilepsy (Gloor et al., 1979) or of an inhibitory, possibly GABAergic activity, as the primary event (Fromm, 1986) remains debatable.

Acknowledgements

Special thanks are given to A. Boehrer for technical assistance.
This work was supported by grants from INSERM (Contrat de Recherche externe n° 866017 and CAR n° 400019) and from "La Fondation pour la Recherche Médicale".

References

Avoli M, Gloor P (1982) Interaction of cortex and thalamus in spike and wave discharges of feline generalized penicillin epilepsy. Exp Neurol 76: 196–217

Avoli M, Gloor P, Kostopoulos G, Gotman J (1983) An analysis of penicillin-induced generalized spike and wave discharges using simultaneous recordings of cortical and thalamic single neurons. J Neurophysiol 50: 819–837

Bures J, Buresova O, Krivanek J (1974) The mechanism and application of Leao's spreading depression of electroencephalographic activity. Publishing House of the Academy of Sciences, Prague, p 399

Depaulis A, Bourguignon JJ, Marescaux C, Vergnes M, Schmitt M, Micheletti G, Warter JM (1988) Effects of gamma-hydroxybutyrate and gamma-butyrolactone derivatives on spontaneous generalized non-convulsive seizures in the rat. Neuropharmacology 27: 683–689

Depaulis A, Vergnes M, Liu Z, Kempf E, Marescaux C (1990) Involvement of the nigral output pathways in the inhibitory control of the substantia nigra over generalized non-convulsive seizure in the rat. Neurosciences 39: 339–349

Di Chiara G, Porceddu ML, Morelli M, Mulas ML, Gessa GL (1979) Evidence for a GABAergic projection from the substantia nigra to the ventromedial thalamus and to the superior colliculus of the rat. Brain Res 176: 273–284

Fromm G (1986) Role of inhibitory mechanisms in staring spells. J Clin Neurophysiol 3: 297–311

Gloor P, Pelligrini A, Kostopoulos GK (1979) Effects of changes in cortical excitability upon the epileptic bursts in generalized penicillin epilepsy of the cat. Electroencephalogr Clin Neurophysiol 46: 274–289

Hunter J, Jasper HH (1949) Effects of thalamic stimulation in unanesthetized animals. EEG Clin Neurophysiol 1: 305–324

Jones EG (1985) The thalamus. Plenum, New York, p 935

Marescaux C, Micheletti G, Vergnes M, Depaulis A, Rumbach L, Warter JM (1984) A model of chronic spontaneous petit mal-like seizures in the rat: comparison with pentylenetetrazol-induced seizures. Epilepsia 25: 326–331

Marescaux C, Vergnes M, Depaulis A, Micheletti G, Warter JM (1992) Neurotransmission in rats' spontaneous generalized nonconvulsive epilepsy. In: Avanzini G, et al (eds) Neurotransmitters in epilepsy. Epilepsy Res [Suppl] (in press)

McLachlan RS, Gloor P, Avoli M (1984) Differential participation of some "specific" and "non-specific" thalamic nuclei in generalized spike and wave discharges of feline generalized penicillin epilepsy. Brain Res 307: 277–287

Mirski MA, Ferrendelli JA (1986) Anterior thalamic mediation of generalized pentylenetetrazol seizures. Brain Res 399: 212–223

Mirski MA, Ferrendelli JA (1987) Interruption of the connections of the mammillary bodies protects against generalized pentylenetetrazol seizures in guinea pigs. J Neurosci 7: 662–670

Paxinos G, Watson C (1982) The rat brain in stereotaxic coordinates. Academic Press, New York

Pellegrini A, Gloor P (1979) Effects of bilateral partial diencephalic lesions on cortical epileptic activity in generalized penicillin epilepsy in the cat. Exp Neurol 66: 285–308

Ross DT, Ebner FF (1990) Thalamic retrograde degeneration following cortical injury: an excitotoxic process? Neuroscience 35: 525–550

Steriade M, Deschenes M (1984) The thalamus as a neuronal oscillator. Brain Res Rev 8: 1–63

Vergnes M, Marescaux C, Micheletti G, Reis J, Depaulis A, Rumbach L, Warter JM (1982) Spontaneous paroxysmal electroclinical patterns in rat: a model of generalized nonconvulsive epilepsy. Neurosci Lett 33: 97–101

Vergnes M, Marescaux C, Micheletti G, Depaulis A, Rumbach L, Warter JM (1984) Enhancement of spike and wave discharges by GABAmimetic drugs in rats with spontaneous petit mal-like epilepsy. Neurosci Lett 44: 91–94

Vergnes M, Marescaux C, Depaulis A, Micheletti G, Warter JM (1987) Spontaneous spike and wave discharges in thalamus and cortex in a rat model of genetic petit mál-like seizures. Exp Neurol 96: 127–136

Vergnes M, Marescaux C, Lannes B, Depaulis A, Micheletti G, Warter JM (1989) Interhemispheric desynchronisation of spontaneous spike-wave discharges by corpus callosum transection in rats with petit mal-like epilepsy. Epilepsy Res 4: 8–13

Vergnes M, Marescaux C, Depaulis A, Micheletti G, Warter JM (1990) Spontaneous spike-and-wave discharges in Wistar rats: a model of genetic generalized nonconvulsive epilepsy. In: Avoli M, Gloor P, Kostopoulos G, Naquet R (eds) Generalized epilepsy: neurobiological approaches. Birkhäuser, Boston, pp 238–253

Vergnes M, Marescaux C, Depaulis A (1990) Mapping of spontaneous spike and wave discharges in Wistar rats with genetic generalized nonconvulsive epilepsy. Brain Res 523: 87–91

Warter JM, Vergnes M, Depaulis A, Tranchant C, Rumbach L, Micheletti G, Marescaux C (1988) Effects of drugs affecting dopaminergic neurotransmission in rats with spontaneous petit mal-like seizures. Neuropharmacology 27: 269–274

Williams D (1953) A study of thalamic and cortical rhythms in petit mal. Brain 76: 50–69

Authors' address: Dr. M. Vergnes, LNBC, Centre de Neurochimie du CNRS, 5, rue Blaise Pascal, F-67084 Strasbourg Cédex, France

J Neural Transm (1992) [Suppl] 35: 85–95

Role of the thalamic reticular nucleus in the generation of rhythmic thalamo-cortical activities subserving spike and waves

G. Avanzini[1], M. de Curtis[1], C. Marescaux[2], F. Panzica[1], R. Spreafico[1], and M. Vergnes[3]

[1] Istituto Neurologico C. Besta, Milano, Italy
[2] Clinique Neurologique de l'Université de Strasbourg, and [3] Centre de Neurochimie du CNRS, Strasbourg, France

Summary. The role of the reticular thalamic nucleus (RTN) in pacing rhythmic cortical activities subserving spike-waves (SW) discharges has been investigated in rats.

Intracellular recordings from thalamic slices in vitro demonstrated that RTN neurons from control animals possess a set of Ca^{2+}/K^+ membrane conductances which enable them to produce rhythmic oscillatory activities.

In vivo, studies of Ca^{2+}-conductance blockade by intrathalamic injections of Cd^{2+} were performed on 24 callosotomized Wistar rats displaying spontaneous SW discharges, bred at the Centre de Neurochimie, Strasbourg. A significant decrement in ipsilateral SW activity was consistently observed in all RTN-injected animals 40 min after Cd^{2+} injection. By contrast, animals which received Cd^{2+} injection into the ventroposterior complex (VP) showed only small changes in ipsilateral SW. It is concluded that Ca^{2+}-dependent oscillatory properties of the RTN are critical for the expression of genetically determined SW discharges in the Wistar model.

Introduction

The role of the thalamic nuclei in pacing rhythmic cortical activities is presently being reconsidered in the light of new emerging information. From a series of experiments carried out in cats by Steriade's group (Steriade and Deschenes, 1984; Steriade et al., 1986, 1987; Mulle et al., 1986), it became increasingly clear that the thalamic reticular nucleus (RTN) is a prime determinant of some rhythmic activities (e.g. sleep spindles).

The present paper reports some recent findings on intrinsic mechanisms responsible for RTN oscillatory properties and on their role in spike-wave (SW) discharge generation in rats.

RTN is a thin laminar nucleus which surrounds the dorsolateral and anterior portions of the dorsal thalamus. It is entirely composed of

GABAergic neurons, from which axons extend to all the other thalamic nuclei. It receives collaterals from both thalamocortical and corticothalamic systems, but does not participate directly in the thalamocortical projection.

As demonstrated for other thalamic nuclei by Deschenes et al. (1984) and by Jansen and Llinas (1984a,b), RTN neurons can fire in two different modes: single-spike/tonic firing or bursting mode, depending on the level of membrane polarization (Kayama et al., 1986; Mulle et al., 1986; Avanzini et al., 1989).

Intracellular recordings from in vivo (Mulle et al., 1986) and in vitro (Avanzini et al., 1989) RTN neurons have shown that in particular conditions bursting discharges tend to recur regularly, giving rise to rhythmic membrane oscillations, which are strictly correlated with EEG spindles (Mulle et al., 1986).

The idea that spindles and SW may share common thalamocortical synchronizing mechanisms is supported by observations in different experimental models of generalized epilepsies (see review in Gloor and Fariello, 1988). It therefore seemed interesting to investigate the putative role of RTN in SW generation in an accredited model of generalized epilepsy: the Wistar rat with SW discharges associated with absence-like seizures (Vergnes et al., 1982).

The experimental design of the in vivo study on epileptic Wistar rats (Avanzini et al., submitted) is based on the results of previous in vitro experiments on RTN physiology in normal rats (Spreafico et al., 1988; Avanzini et al., 1989; de Curtis et al., 1989).

Methods and results of both in vitro and in vivo studies are reported and discussed below, with reference to the involvement of RTN in SW generation.

Methods

In vitro experiments

These experiments were carried out at the Laboratory of Neurophysiology in Milan on 200–350 g adult rats. The animals were decapitated by guillotine under light ether anaesthesia. The brain was quickly removed and cut along the midline with a razorblade. A small block of brain tissue containing the thalamus was dissected, glued by its dorsal surface to a brass holder, and immersed in cold (10°C) oxygenated artificial cerebrospinal fluid (ACSF: NaCl 134 mM, KCl 5 mM, Mg_2SO_4 + H_2O 2 mM, KH_2PO_3 1.25 mM, $CaCl_2$ + $2H_2O$ 2 mM, $NaHCO_3$ 16 mM, glucose 10 mM). Horizontal thalamic slices (300–400 µm thick) were cut from the ventral surface of the specimen by means of a Vibratome. Slices in which the RTN and the ventroposterior complex (VP) were recognizable under the stereomicroscope were transferred to the incubation chamber, where they were perfused with oxygenated ACSF at 36–37°C and exposed to a humidified atmosphere of 95% O_2 and 5% CO_2. The slices were incubated for about 1 h before the start of the experiment.

Monopolar tungsten electrodes were placed under microscopic control into the VP and into the internal capsule (IC) for subsequent activation of the RTN or VP neurons.

Glass micropipettes for intracellular recording filled with 4 M K-acetate or 3 M KCl or 5% horseradish peroxidase (HRP Boering Grade I) solution in 0.2 M KCl, were positioned in RTN, or in VP.

The resistance was about 80–120 MΩ for the K-acetate and KCl electrodes and 100–150 MΩ for the HRP-filled electrodes.

The membrane potentials were recorded with a Neurodata Instrument Corp. (USA) pre-amplifier equipped with a Wheatstone bridge utilized to perform intra-cellular current injections.

Input-resistance values were calculated on the initial linear portion of current-voltage curves produced by measuring the changes in membrane voltage induced by 80–100 ms hyperpolarizing current pulses applied through the recording micro-electrodes. The time-constant values were calculated by identifying an exponential function derived from the application of a least-squares criterion to the logarithms of the data points extrapolated by the membrane deflection induced by injections of 0.2–0.4 nA current pulses.

The data were recorded on magnetic tape (Racal FM4D) and analysed offline by means of a Digital PDP 11/34 computer system, together with a Tektronix 5110 digital oscilloscope and a chart recorder (Linseis, Type 2045). Only data from those neurons producing prolonged (>10 min), stable recordings and with a membrane potential over −50 mV were accepted for analysis.

When 1 mM cadmium (Cd^{2+}) was added to the solution, phosphate sulphate and $CaCl_2$ were omitted, and $MgSO_4$ was replaced with $MgCl_2$. During some experiments, pharmacological agents were added to the perfusion fluid without adjusting the tonicity: 1 µM tetrodotoxin (TTX) (n = 3); 20 mM tetraethylammonium (TEA) (n = 4); 100 µM apamine (n = 3); 1 µM 8-bromo adenosine 3′,5′-cyclic monophosphate (8-Br cyclic AMP) (n = 4).

The following excitatory amino acids (EAAs) and their receptor agonists, diluted in a 150-mM NaCl solution (pH 8), were iontophoretically tested by utilizing a multibarrel pipette; glutamate (1 and 0.5 M); aspartate (1 and 0.5 M); N-methyl-D-aspartate (NMDA, 20 and 50 mM); kainate (100 mM); quisqualate (100 mM). The drugs were retained in the pipette by a slight negative current (2–4 nA). The selective NMDA-receptor antagonist 2-amino 5-phosphonovalerate (APV) was applied to the preparation in superfusion (50–100 µm); its effect showed up 10 min after the superfusion medium change. All compounds were supplied by Sigma.

Continuous evaluation of the input resistance during drug ejections was performed by measuring the potential deflection induced by intracellular hyperpolarizing current pulse injections. The amplitude and duration of the hyperpolarizing current pulse varied from one experiment to another, between 0.1 and 0.4 nA and 100 and 150 ms, respectively, but were fixed within the same experiment.

In vivo experiments

These were carried out at the Centre de Neurochimie in Strasbourg on 20 Wistar rats (350–550 g) from the sixth to seventh generation of the strain inbred for occurrence of SW (Vergnes et al., 1982) according to the current procedure employed in that laboratory (Depaulis et al., 1988; Vergnes et al., 1989). Under pentobarbitone anaesthesia (40 mg/kg i.p.) rats were fixed with earbars in a stereotaxic apparatus and the corpus callosum was sectioned with a surgical blade stereotaxically guided along the midline. Guide cannulae (0.4 mm outside diameter and 0.3 mm inside diameter) were stereotaxically implanted. Up to 4 cannulae were implanted in each animal, directed toward the uppermost borders of RTN and VP.

Finally, four single contact electrodes made from stainless-steel dental screws were inserted in the skull on the frontal and parietal regions of both sides. The screws were

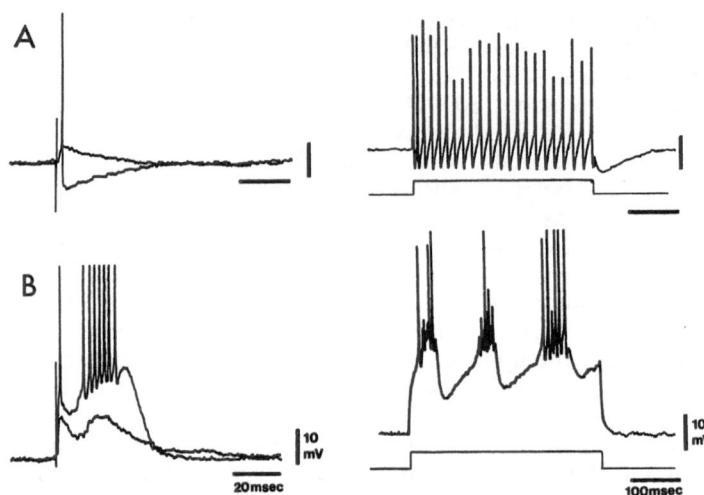

Fig. 1. In vitro intracellular recording of an RTN neuron from a normal rat: **A** at resting membrane potential (−58 mV); **B** during membrane hyperpolarization (−78 mV) induced by steady current injection. *Left:* superimposition of postsynaptic responses evoked by IC stimulations at two different intensities (infra- and supra-threshold for action potentials). *Right:* effect of intracellular injection of long-lasting depolarizing current pulses (from Avanzini et al., 1990, with permission)

connected to a microconnector secured to the skull with stainless-steel anchoring screws and acrylic resin.

After recovery from anaesthesia, the baseline EEG from freely moving animals was recorded for several days to characterize individual SW discharge rates. Finally, the effect of 0.5–1 μl of Cd^{2+} 1 mM solution on ipsilateral SW was tested by local injection. Injection cannulae (0.28 mm outside diameter, 0.18 mm inside diameter) connected with a microsyringe and inserted in the pre-implanted guide canula were employed. The injection procedure did not require anaesthesia and took altogether no more than 5 min, thus allowing the immediate evaluation of the acute Cd^{2+} effect on SW discharges.

Tracks of guide cannulae and of protruding injections cannulae were easily reconstructed on histological sections after completion of the recording sessions.

Results

Rhythmic bursting discharge of in vitro RTN neurons

Burst firing was physiologically observed in RTN neurons at membrane potential values below −60 mV (Fig. 1). When RTN neurons were stimulated directly with long-lasting (600–1000 ms) and large (0.3–1 nA) depolarizing current pulses from a hyperpolarized level, repetitive 6–8 Hz bursting discharges were elicited (Fig. 1). Each burst was followed by a pronounced hyperpolarization (BAHP), lasting 80–120 ms (Fig. 1) and resulting in a rhythmic oscillatory behaviour of the membrane potential. Perfusion with

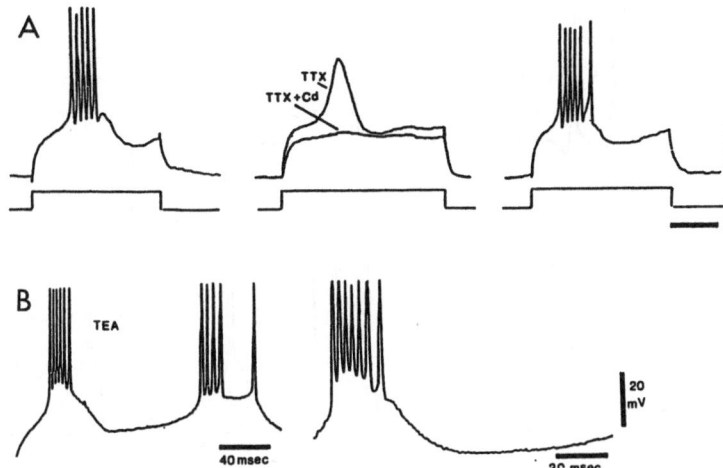

Fig. 2. A *left:* bursting response evoked by a depolarizing current pulse in a RTN neuron recorded in vitro from a normal rat. *Centre:* blockade of Na$^+$ spikes with 1 µM TTX uncovers a Cd^{2+} (1 mM)-sensitive Ca^{2+} spike. *Right:* recovery of the fully developed burst after washing. **B** perfusion with TEA (20 mM) impairs the early phase of BAHP and prolongs the late slow-decaying phase

tetrodotoxin (TTX) 1 µM abolished Na$^+$ spikes, thus uncovering a slow all-or-none depolarizing potential underlying the burst, which was in turn abolished by Cd^{2+} 1 mM (Fig. 2), demonstrating its Ca^{2+}-dependency.

Perfusion with tetraethylammonium (TEA) 20 mM demonstrated the bimodal nature of BAHP, which consisted of an initial TEA-sensitive fast phase peaking at about 20 ms and a slower TEA-insensitive decay phase, which was not reduced but rather enhanced by TEA perfusion (Fig. 2).

On the basis of these pharmacological tests, the burst-BAHP sequence was attributed to a set of Ca^{2+}/K$^+$ membrane conductances responsible for a low-threshold Ca^{2+} inward current underlying bursts and for a fast and slow Ca^{2+}-dependent K$^+$(K$^+_{(Ca^{2+})}$) outward current, which BAHPs are due to. The apparent facilitation of the slow K$_{(Ca^{2+})}$ during TEA perfusion was explained by a TEA-induced increase in Ca^{2+} influx due to the broadening of the spike repolarization.

The existence of a well-developed biphasic BAHP was the main physiological difference between RTN and VP neurons. The latter consistently lacked long-lasting K$^+$-dependent after-hyperpolarizations and never showed a tendency to produce repetitive bursting discharges.

The effects of the following drugs on oscillatory properties of RTN neurons were tested (Fig. 3). Cadmium 1 mM, blocking K$^+_{(Ca^{2+})}$ conductances before a total Ca^{2+} blockade, suppressed the repetitive burst discharges. Apamine 100 mM, which blocks slow K$^+_{(Ca^{2+})}$ conductances, did the same. By contrast, TEA 20 mM, which suppresses the early fast K$^+_{(Ca^{2+})}$ current with a reciprocal enhancement of the late BAHP component, facilitated and prolonged the rhythmic discharges. Note also that

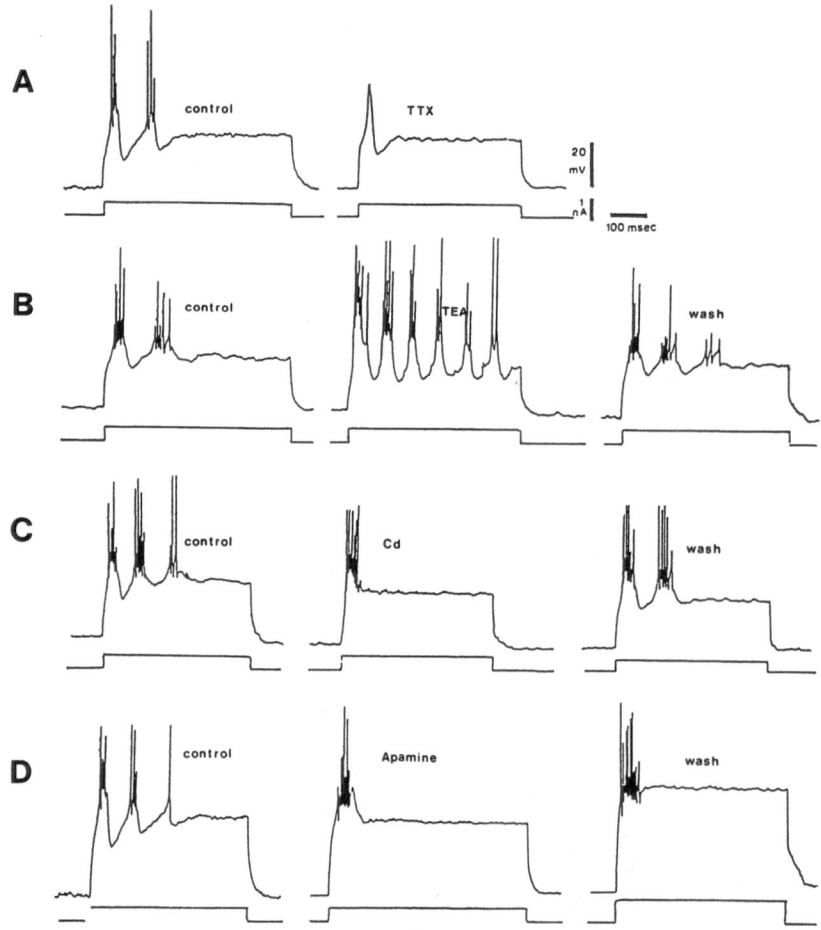

Fig. 3. Effects of TTX 1 μM (**A**), TEA 20 mM (**B**), Cd^{2+} 1 mM (**C**) and Apamine 100 μM (**D**) on rhythmic bursting behaviour of different RTN neurons recorded in vitro from normal rats (from Avanzini et al., 1989, with permission)

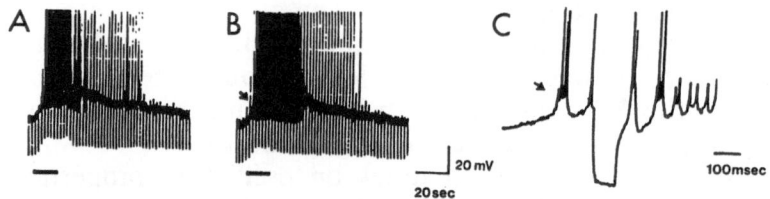

Fig. 4. Effects of 1 M glutamate (**A**) and 1 M aspartate (**B**) iontophoretic application on an RTN neuron recorded in vitro from a normal rat. The ejection time is marked by a horizontal bar below the traces. **C** fast recording of the response depicted in B (arrow). Membrane resistance was continuously monitored by 0.5 nA hyper-polarizing current pulses, resulting in negatively trending vertical bars in A and B. The membrane deflection induced by one such pulse is better seen in C

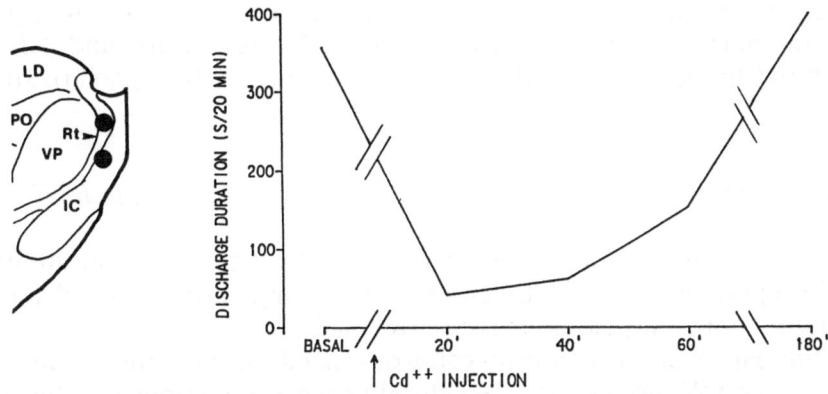

Fig. 5. Reversible antagonistic effect of unilateral Cd^{2+} injection in RTN on SW in a callosotomized Wistar rat from the epileptic strain studied in vivo. Changes in SW generation ipsilateral to the injection are expressed as total discharge duration (s) over 20 min (from Avanzini et al., submitted, with permission)

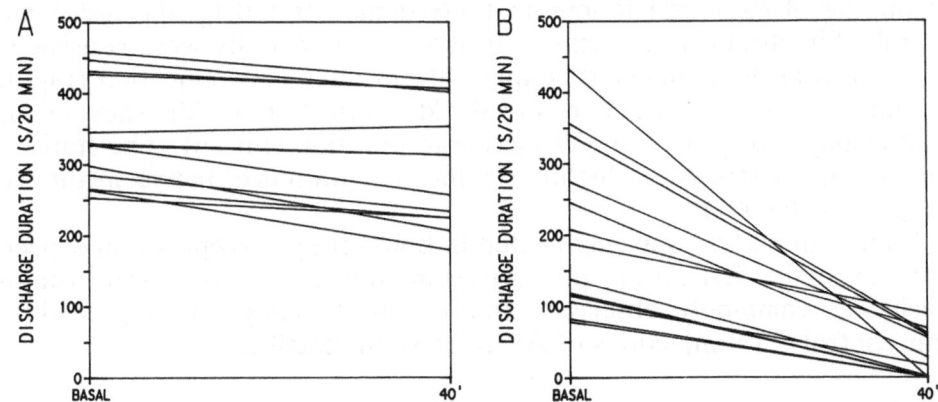

Fig. 6. Durations (s) of unilateral SW over 20 min before and 20–40 min after ipsilateral Cd^{2+} injection in VP (**A**) and RTN (**B**) callosotomized rats. In vivo recordings from two groups of Wistar rats of the epileptic strain (from Avanzini et al., submitted, with permission)

the rhythmic behaviour was likewise abolished by TTX 1 mM, in spite of a well-developed BAHP following the low-threshold Ca^{2+} spike uncovered by Na$^+$ blockade. This could be a consequence of the abolition of Ca^{2+} influx which occurs during Na$^+$ spike. Alternatively, it could be attributable to the block of a TTX-sensitive, regenerative Na$^+$ current activated during the burst generation.

Iontophoretic application of glutamate (0.5 and 1M), aspartate (0.5–1M), NMDA (20–50 mM), kainate (100 mM) and quisqualate (100 mM) induced membrane depolarization with sustained cell firing. Aspartate and NMDA were particularly effective in inducing prolonged discharges, which in some neurons took the form of a burst firing associated with an apparent increase in membrane input-resistance (Fig. 4). Aspartate and NMDA-induced burst firing was facilitated by slight membrane hyperpolarization.

In vivo blockade of SW by Ca^{2+} manipulation procedures in RTN

The effect of local administration of Cd^{2+} 1 mM was tested on 20 Wistar rats of the epileptic strain. Altogether 13 Cd^{2+} injections in RTN and VP and 4 ACSF control injections were performed.

The animals were previously callosotomized to prevent commissural transmission of SW discharges. Two closely spaced injections of $0.25-0.5\,\mu l$ of Cd^{2+} 1 mM in RTN consistently induced an ipsilateral long-lasting inhibition of SW in 13 instances.

A representative example of a reversible Cd^{2+}-induced SW suppression is shown in Fig. 5. In other cases, full recovery of the SW discharge duration to the pre-injection level was not obtained, which is consistent with the finding of structural lesions in RTN demonstrated by the histological controls. Significant decrements in ipsilateral SW activity were consistently observed in all RTN-injected animals 40 min after Cd^{2+} injection (Fig. 6). By contrast, animals which received Cd^{2+} injection in VP showed only small changes in ipsilateral SW discharge duration (Fig. 6). The statistical analysis demonstrated a highly significant difference between the two groups ($p < 0.0001$).

Contralateral SW were unaffected in both groups except for an aspecific decrease in the first 10 min post-injection (due to the increased arousal), which was commonly observed after different manipulation procedures. Four control RTN injections of ACSF were ineffective.

Discussion

The present results obtained by pharmacological manipulations of RTN functional activity are consistent with the findings made in lesion experiments (Vergnes et al., 1990, this volume) in supporting a critical role of RTN in the generation of SW discharges in Wistar rats.

Since RTN does not participate in thalamocortical projection, its influence on the cerebral cortex must be mediated by thalamocortical (TC) neurons which are the only target of RTN efferent projections (Jones, 1975). In fact, large electrolytic lesions involving most of the thalamocortical relay nuclei were found to impair ipsilateral SW discharges in Wistar rats (Vergnes et al., 1990). Owing to the massive character of the lesions, however, it was difficult to give a proper account of the relative role of TC, versus RTN neuronal lesion in inducing this effect. More recently, the experiments were repeated by local injection of the cytotoxic agent ibotenic acid to induce restricted focal thalamic lesions not contaminated by lesions of fibres "en passage" (Vergnes et al., 1992; Avanzini et al., submitted). It was thus demonstrated that, while SW were consistently abolished in RTN-lesioned rats, lesions restricted to individual nuclei belonging to the TC system (i.e. VP) had no effect on SW discharge.

The main question addressed in our experiments concerned the specific properties of RTN which could account for its patho-physiological role in SW discharge generation.

The suppression of rhythmic bursting discharges in RT neurons in vitro by Cd^{2+} was paralleled by its antagonism of SW in vivo when injected in RTN. We therefore concluded that SW in Wistar rats are controlled by RTN through a Ca^{2+}-dependent mechanism. Moreover, since rhythmic bursting activity of RTN neurons is specifically dependent on the voltage-dependent low-threshold Ca^{2+} current (T current), it may logically be assumed that the T current is primarily involved in RTN control of SW discharges.

It must be remembered, however, that the T current is common to all TC neurons (Deschenes et al., 1984; Jansen and Llinas, 1984a,b), which had never been found capable of producing self-sustained oscillatory discharges in in vitro experiments (Jansen and Llinas, 1984a,b; Spreafico et al., 1988).

According to the results of our in vitro experiments, a key role in determining repetitive discharges in RTN neurons has to be ascribed to Ca^{2+}-dependent $K^{+}_{(Ca^{2+})}$ conductance, which is coactived with low-threshold Ca^{2+} conductance during bursting activity. The $K^{+}_{(Ca^{2+})}$ conductance is responsible for the post-burst hyperpolarization (BAHP) which drives the membrane potential towards the level at which the low-threshold Ca^{2+} conductance is active, thus allowing the reproduction of the next Ca^{2+}-dependent burst in a sequence. It is noteworthy that the only other thalamic structure that has been found to be provided with a similar $K^{+}_{(Ca^{2+})}$ conductance, the paratenial nucleus (McCormick and Prince, 1988), also displays oscillatory rhythmic discharges.

Different synaptic systems impinging on RTN may be important in triggering the intrinsic oscillatory mechanisms. Cholinergic (Ben Ari et al., 1976; McCormick and Prince, 1986) and adrenergic (McCormick and Prince, 1988) brain stem afferences are known to modulate RTN activity in a state-dependent mode. In addition, our experiments (see also de Curtis et al., 1989) demonstrated that cortical descending influences acting on RTN are mediated by EAA and that aspartatergic cortical projections act selectively on NMDA receptors. The effectiveness of antagonists of EAA in reducing SW discharges in Wistar rats (Marescaux et al., in press) suggests that the cortical input to RTN may have a part in SW generation. On the other hand, the precipitating effect of GABAergic drugs on SW discharge in Wistar rats (Vergnes et al., 1984; Marescaux et al., in press) stresses once again the pathophysiological importance of GABAergic circuitry, of which RTN is a main part.

Further experiments are required to ascertain whether RTN is the site of
the primary genetically determined defect responsible for SW in the Wistar
rat, or whether it is only secondarily involved in pathophysiological mechanisms arising elsewhere.

Acknowledgements

This work was partially supported by the Paolo Zorzi Foundation for Neurosciences.
We thank M. T. Pasquali for her assistance in editing the text.
The collaboration is supported by a joint INSERM — CNR grant.

References

Avanzini G, de Curtis M, Panzica F, Spreafico R (1989) Intrinsic properties of nucleus reticularis thalami neurones of the rat studied in vitro. J Physiol 416: 111–122
Avanzini G, de Curtis M, Spreafico R (1992) Physiological properties of GABAergic thalamic reticular neurons studied in vitro. Relevance to thalamo-cortical synchronizing mechanisms. In: Avanzini G, Fariello R, Heinemann U, Engel J (eds) Neurotransmitters in epilepsy. Elsevier, Amsterdam (in press)
Avanzini G, Vergnes M, Spreafico R, Marescaux C (1992) Calcium dependent regulation of genetically determined spike and waves by the reticular thalamic nucleus of rats. Epilepsia (submitted)
Ben Ari Y, Dingledine R, Kanazawa I, Kelly JS (1976) Inhibitory effects of acetylcholine on neurons in the feline nucleus reticularis thalami. J Physiol 261: 647–671
de Curtis M, Spreafico R, Avanzini G (1989) Excitatory amino acids mediate responses elicited in vitro by stimulation of cortical afferents to reticularis thalami neurons of the rat. Neuroscience 33: 275–283
Depaulis A, Vergnes M, Marescaux C, Lannes B, Warter JM (1988) Evidence that activation of GABA receptors in the substantia nigra suppresses spontaneous spike-and-wave discharges in the rat. Brain Res 448: 20–29
Deschenes M, Paradis M, Roy JP, Steriade M (1984) Electro-physiology of neurons of lateral thalamic nuclei in cat: resting properties and burst discharges. J Neurophysiol 51: 1196–1219
Gloor P, Fariello R (1988) Generalized epilepsy: some of its cellular mechanisms differ from those of focal epilepsy. Trends Neurosci 11: 63–68
Jansen H, Llinas R (1984a) Electrophysiological properties of guinea-pig thalamic neurones: an in vitro study. J Physiol 349: 205–226
Jansen H, Llinas R (1984b) Ionic basis for the electro-responsiveness and oscillatory properties of guinea-pig thalamic neurones in vitro. J Physiol 349: 227–247
Jones EG (1975) Some aspects of the organization of the thalamic reticular complex. J Comp Neurol 162: 285–308
Kayama Y, Sumimoto I, Ogawa T (1986) Does the ascending cholinergic projection inhibit or excite neurons in the rat thalamic reticular nucleus? J Neurophysiol 56: 1310–1320
Marescaux C, Vergnes M, Depaulis A, Micheletti G, Warter JM (1992) Neurotransmission in rats' spontaneous generalized non convulsive epilepsy. In: Avanzini G, Fariello R, Heinemann U, Engel J (eds) Neurotransmitters in epilepsy. Elsevier, Amsterdam (in press)

McCormick DA, Prince DA (1986) Acteylcholine induces burst firing in thalamic reticular neurons by activating a potassuim conductance. Nature 319: 402–405

McCormick DA, Prince DA (1988) Noradrenergic modulation of firing pattern in guinea pig and thalamic neurons, in vitro. J Neurophysiol 59: 978–996

Mulle C, Madariaga A, Deschenes M (1986) Morphology and electro-physiological properties of reticularis thalami neurons in cat: in vivo study of a thalamic pacemaker. J Neurosci 6: 2134–2145

Spreafico R, de Curtis M, Frassoni C, Avanzini G (1988) Electro-physiological characteristics of morphologically identified reticular thalamic neurons from rat slices. Neuroscience 27: 629–638

Steriade M, Deschenes M (1984) The thalamus as a neuronal oscillator. Brain Res Rev 8: 1–62

Steriade M, Dominich L, Oakson G (1986) Reticularis thalami neurons revisited: activity changes during shifts in state of vigilance. J Neurosci 6: 68–81

Steriade M, Dominich L, Oakson G, Deschenes M (1987) The deafferented reticularis thalami nucleus generates spindles rhythmicity. J Neurophysiol 57: 260–273

Vergnes M, Marescaux C (1992) Cortical and thalamic lesions in rats with genetic absence epilepsy (this volume)

Vergnes M, Marescaux C, Micheletti G, Reis J, Depaulis A, Rumbach L, Warter JM (1982) Spontaneous paroxysmal electroclinical patterns in rat: a model of generalized non-convulsive epilepsy. Neurosci Lett 33: 97–101

Vergnes M, Marescaux C, Micheletti G, Depaulis A, Rumbach L, Warter JM (1984) Enhancement of spike and wave discharges by GABA-mimetic drugs in rats with spontaneous petit mal-like epilepsy. Neurosci Lett 44: 91–94

Vergnes M, Marescaux C, Lannes B, Depaulis A, Micheletti G, Warter JM (1989) Interhemispheric desynchronization of spontaneous spike-wave discharges by corpus callosum transection in rats with petit mal-like epilepsy. Epilepsy Res 4: 8–13

Vergnes M, Marescaux C, Depaulis, A, Micheletti G, Warter JM (1990) Spontaneous spike-and-wave discharges in Wistar rats: a model of genetic generalized convulsive epilepsy. In: Avoli M, Gloor P, Kostopoulos G, Naquet R (eds) Generalized epilepsy. Neurobiological approaches. Birkhäuser, Boston, pp 238–253

Authors' address: Prof. G. Avanzini, MD, Laboratory of Neurophysiology, Istituto Neurologico C. Besta, Via Celoria 11, I-20133 Milano, Italy

J Neural Transm (1992) [Suppl] 35: 97–108

Responses to N-methyl-D-aspartate are enhanced in rats with Petit Mal-like seizures

R. Pumain[1], J. Louvel[1], M. Gastard[1], I. Kurcewicz[1], and M. Vergnes[2]

[1] Unité de Recherches sur l'Epilepsie, INSERM U 97, Paris, and [2] Département de Neurophysiologie et de Biologie des Comportements, Centre de Neurochimie du CNRS, Strasbourg, France

Summary. The responses to the glutamate agonist N-methyl-D-aspartate (NMDA) were studied in the sensori-motor cortex of rats with petit mal-like seizures. In a first study, the changes in extracellular concentration of calcium elicited through ionophoretic application of NMDA at various depths in the cortex were measured in vivo. The results show that in the cortex of epileptic rats the NMDA responses are much more widely distributed than in the cortex of control rats. In a second study, a current-source density analysis of the responses elicited through electrical stimulation of the white matter was performed in slices of neocortex in vitro. These findings show that the NMDA-dependent component of the synaptic responses are more widely distributed and of longer duration in the cortex of epileptic rats than in that of control rats. Taken together, these results suggest that in this model of absence epilepsy NMDA-dependent mechanisms are important in the triggering and maintenance of epileptic activity.

Introduction

Although a number of experimental models of generalized epileptiform discharges are available (Avoli et al., 1990), there is a scarcity of chronic animal models of Petit Mal seizures. Therefore, it has been difficult to assess whether the mechanisms underlying epileptogenesis in this type of seizures were similar to those described for other types of epilepsies. Recently, M. Vergnes and her colleagues in Strasbourg have been able to select a strain of Wistar rats displaying spontaneous generalized, non-convulsive epileptic seizures (Marescaux et al., 1984; Vergnes et al., 1982). The evidence for concluding that such rats were representative of a model of Petit Mal seizures was based on behavioral, electrophysiological and pharmacological data (Marescaux et al., 1984; Vergnes et al., 1987, 1990).

Concomitantly, a strain of control rats, who never displayed spike-and-wave discharges, was selected in the same laboratory.

We had previously shown that, in animals bearing a cortical epileptogenic chronic focus, the responses to N-methyl-D-aspartate (NMDA), an agonist of the excitatory neurotransmitter glutamate, were more widely distributed than in control animals (Pumain et al., 1986). Due to the peculiar properties of the ionic channels associated with NMDA, namely the voltage-dependence and the high permeability to calcium ions (Ascher and Nowak, 1988b; MacDermott et al., 1986; Mayer and Westbrook, 1987, 1988; Mayer et al., 1984; Nowak et al., 1984; Pumain et al., 1987), to the fact that the enhancement of the NMDA-mediated transmission could produce seizure-like activity in slices of various nervous structures (Anderson et al., 1986; Avoli et al., 1987; Stanton et al., 1987; Walther et al., 1986), and that antagonists of NMDA receptors displayed antiepileptic properties (Croucher et al., 1982; Meldrum et al., 1983), it has been suggested that the wide distribution of these receptors observed in epileptogenic zones could be an important factor in the triggering and maintenance of epileptic activity (Heinemann et al., 1986; Pumain et al., 1986). Furthermore, in more recent studies, we have observed similar findings in pieces of cortical epileptogenic tissue obtained from patients undergoing neurosurgery to be relieved from intractable epilepsy (Louvel et al., 1992). Therefore, we tested whether such a feature, observed in various chronic epilepsies, could also be observed in this model of Petit Mal absence. If this assumption is verified, then a common mechanism may underlie epileptogenesis in both focal epilepsies, arising from zones with demonstrable lesions, and in non-lesional, generalized epilepsies. Since in this strain of epileptic rats spike-and-wave discharges are most prominent in the neocortex and lateral thalamus (Vergnes et al., 1982, 1990), the recordings in the present study were performed in the sensori-motor cortex.

In order to determine the extension and location of NMDA responses in this model, two different methods were employed. The first method was the measurement of ionic responses in vivo, using ion-selective microelectrodes, during ionophoretic applications of excitatory amino acids: it has been shown that the fall, in the extracellular space, of the concentration of calcium or sodium ions during application of these substances is mainly due to a direct action on the corresponding receptors and to the permeability changes thus induced in postsynaptic membranes (Pumain and Heinemann, 1985; Pumain et al., 1987). Therefore, the measurement of the extracellular ionic changes induced through local, constant applications at various depth under the cortical surface of these substances allows to determine the zones where the responses are the largest. The other agonists of glutamate, quisqualate and kainate, acting on different receptors were also tested for control. The second method was the measurement of synaptic responses evoked in slices of neocortex maintained in vitro, during electrical stimulation of the white matter, using the current-source density analysis, in presence or in absence of NMDA antagonists.

Materials and methods

The experiments were performed on 16 adult male Wistar rats from the Strasbourg strains, weighing 250 to 400 gm, 12 for the experiments in vivo (6 control rats and 6 epileptic rats), the other rats being used for preparing slices of neocortex. (3 epileptic rats and one control).

For the experiments in vivo, the animals were anesthetized through inhalation of a mixture of oxygen (2/3) and nitrous oxide (1/3), in which fluothane was added (2% to 4%). The electrocardiogram was monitored throughout the experiment and the temperature was kept at 36°–37°C with a heating blanket. The animal was placed in a stereotaxic frame, a craniectomy was performed and the cisterna magna was opened. Once the cerebral cortex was exposed, it was covered with an agar-agar gel to prevent drying. Two electrocorticographic electrodes were positioned on the frontal cortex, one on each side, and both bipolar and monopolar recordings were performed.

The selective microelectrodes were made according to the method of Lux and Neher (1973) as described previously (Pumain and Heinemann, 1985). In brief, the tip of two-barrelled micropipettes were broken back to obtain a tip diameter of about $2 \mu m$. One channel was filled with a 100 mM solution of chloride salt of the ion to be measured (sodium or calcium), and a column of about $200 \mu m$ of the selective resin (Ammann, 1986) was sucked back in the corresponding tip after silanization. The other reference channel was filled with a solution containing 150 mM NaCl. The selective electrodes were retained if their slope was over 25 mV for the calcium electrodes and over 52 mV for the sodium electrodes. The electrodes were calibrated again at the end of the experiment.

A five-channel ionophoretic micropipette was glued to the selective electrode such that the tips of the two pipettes were distant by no more than $10 \mu m$. The electrode assembly was fixed to the headstage of an electrometer with a very high input impedance (Jens Mayer) and the channels of the ionophoretic electrode were connected to a Neurophore (Med. Sys. Corp.). The assembly was then lowered in the sensori-motor cortex of the animal by steps of $200 \mu m$, from surface to a depth of $2,200 \mu m$, perpendicular to the cortical surface. The solutions contained in the channels of the ionophoretic pipettes were the following: sodium L-glutamate (Glu), 0.2 M, pH 7.2; sodium N-methyl-D-aspartate, 0.2 M, pH 7.2 M; sodium quisqualate (Quis), 0.1 M, pH 7.2; sodium kainate (Ka), 0.05 m, pH 7.2. A retaining current of 15 nA was applied to all channels.

Concerning the in vitro experiments, the cortical slices were prepared as previously described (Abbes et al., 1990). The brain was rapidly removed from the decapitated and craniotomized rats, immersed in cold artificial cerebro-spinal fluid (ACSF) and trimmed. Slices were prepared from the sensori-motor cortex using a vibroslicer (Campden) and transferred to a stocking chamber and later to a modified Andersen chamber for recording. The ACSF was warmed at 35–36°C and the flow had an approximate rate of 2 ml/mn. Its composition was, in mM: Na^+, 151.25; Cl^-, 133; K^+, 5; Mg^{2+}, 2; Ca^{2+}, 2; H_2PO_4, 1.25; SO_4^{2-}, 2; $NaH CO_3^-$, 26; Glucose, 10. The pH was adjusted at 7.4. The antagonist of NMDA, 2-amino-phosphono-valerate (APV), was sometimes added to the bath at a concentration of $25 \mu M$.

The stimulations of the white matter were delivered through a bipolar tungsten electrode (tip diameters $100 \mu m$, inter-tip distance $500 \mu m$), every 5 sec, placed between 1,800 and $2,500 \mu m$ below the cortical surface. The stimuli were bipolar pulses of $50–150 \mu sec$ duration, whose intensities were fixed arbitrarily at a value which produced, at $250–350 \mu m$ below the cortical surface, a response of 2 to 4 mV in amplitude. In such conditions, the stimulus intensities were in the range of $50–350 \mu A$. The stimulus intensity was kept constant for a given potential profile. The field potentials were recorded (band pass DC-3000 Hz) using glass micropipettes filled with a

150 mM NaCl solution (impedance 5 to 10 MΩ). The reference electrode was an Ag/AgCl agar bridge.

The current-source density method

The theory of current-source density (CSD) analysis has been described in detail in many good reviews (Freeman and Nicholson, 1975; Mitzdorf, 1985; Mitzdorf and Singer, 1978, 1979; Nicholson and Freeman, 1975; Richardson et al., 1987). Suffice to say that if the potentials evoked through single electrical shocks are measured vertically through the cortical layers at close positions, and assuming that the conductivity in the extracellular space does not change significantly with cortical depth, then the current density in the extracellular space can be expressed as the second spatial derivative of the potential. The method requires that the potentials are identical with successive stimulations (stimulations at low frequency) and that the distance between two successive recording sites is small. It is then possible to localize the sinks and sources of current resulting from the activation of populations of neurons in a given structure. In the present study, this step value was 50 µm and at each recording site, the value of the potential was in fact the average of 8 successive potential responses. Furthermore, we substituted for the potential value a symmetric weighted and averaged value computed from the values recorded from three successive recording sites (Freeman and Nicholson, 1975) in order to reduce the non-physiological noise which results from the application of the derivatives equation. Therefore the CSD analysis started two steps below the first position recorded and ended two steps above the last one. The field potentials were stored on tape and later digitized, averaged and a process of reduction of the stimulus artefact was sometimes used. The CSD values were then computed using a home-made program (J. Louvel).

Results

Experiments in vivo

Base-line levels of extracellular free calcium ($[Ca^{2+}]_0$) and sodium ($[Na^+]_0$) were determined to be between 1.2 and 1.3 mM and between 140 and 148 mM respectively. No significant difference was found between control and epileptic rats.

NMDA was applied every 200 µm from the cortical surface to the white matter, usually with an ionophoretic current of 20 nA. In the control rats, NMDA produced large decreases in $[Ca^{2+}]_0$ mainly in the superficial layers, the responses decaying steadily towards the white matter, as shown in Fig. 1A. To minimize possible problems due to different transport number of the ionophoretic electrodes, or to variable intertip distances between the ionophoretic pipettes and selective electrodes, the responses are normalized, each point representing the mean of percentages of the maximal $[Ca^{2+}]_0$ changes recorded in each electrode track. In this laminar profile, each point is the mean of six values, the horizontal bars indicating the S.E.M. In the epileptic rats, we also recorded large responses in the superficial layers, but in contrast to the control animals, the responses were

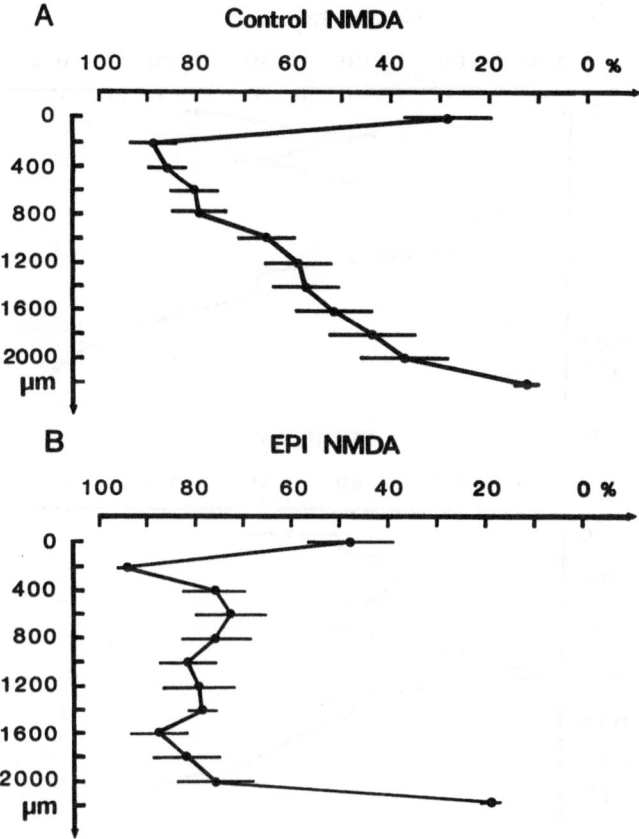

Fig. 1. Laminar profiles of the $[Ca^{2+}]_0$ decreases induced through ionophoresis of N-methyl-D-aspartate (NMDA) in the sensory motor cortex of control (**A**) and of epileptic (**B**) rats, from surface to white matter. The laminar profiles are expressed at each depth as the mean (n = 6) of the percentage of the maximum $[Ca^{2+}]_0$ changes for each electrode track. Horizontal bars represent ±SEM

as large in the deep layers (Fig. 1B), indicating that in the epileptic rats the NMDA receptors are widely distributed throughout the cortex. The maximal absolute values, although difficult to compare for the reasons mentioned above, do appear to be similar for control and epileptic rats.

In order to determine whether the difference observed was or not specific for NMDA responses, we also tested the responses to the application of the other agonists of glutamate, Quis and Ka. For these amino acids, it is more convenient to measure the decreases in $[Na^+]_0$, since the corresponding ionic channels appear to be very little permeable to calcium (Ascher and Nowak, 1988a). In Fig. 2 are shown the laminar profiles obtained for Ka: the zones at which the largest responses were observed were the middle layers in both the control and epileptic rats, although the depth of the peak responses are not exactly the same (Fig. 2A, B). In

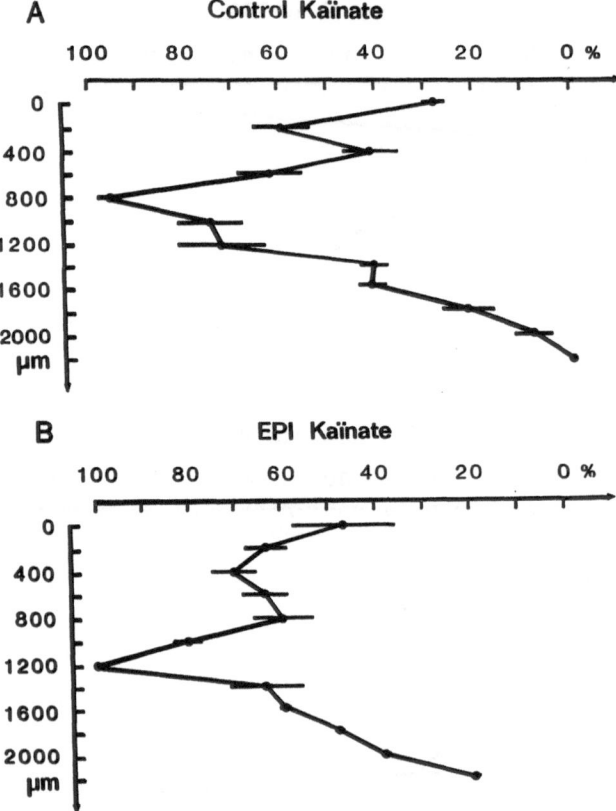

Fig. 2. Laminar profiles of the $[Na^+]_0$ decreases induced through ionophoresis of Kainate in the sensory motor cortex of control (**A**) and of epileptic rats (**B**), from surface to white matter. The laminar profiles are expressed at each depth as the mean (n = 4) of the percentage of the maximum $[Na^+]_0$ for each electrode track. Horizontal bars represent ±SEM

these experiments, each point is the mean of 4 values. Similar results were obtained with Quis. These findings suggest that the distribution of NMDA receptors is much more profoundly altered in the epileptic animals than that of the Quis and Ka receptors, when comparing to the control animals.

Experiments in vitro

In Fig. 3 and 4 are shown the CSD analysis of responses taken from a cortical slice of a control and of an epileptic rat respectively. At the left part of the figures are displayed the laminar potential responses recorded every 50 μm from the surface to the white matter, during regular and constant stimulations of the white matter. At the right part are shown the corresponding CSD profiles. The current sinks, which may be considered to

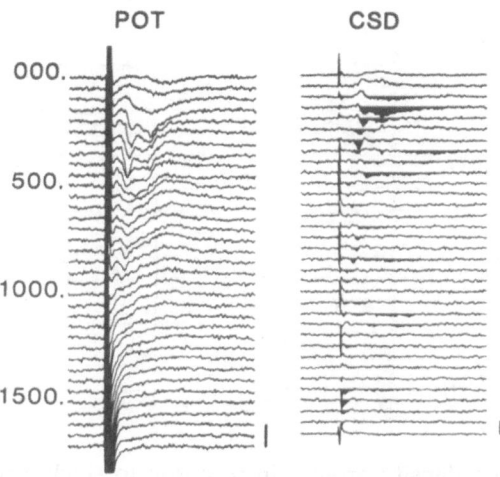

Fig. 3. Laminar potential (POT) and current source density (CSD) profiles in response to single shock stimulations of the white matter in slices of sensory motor cortex of a control rat. The recording electrode was displaced by steps of 50 µm. the slice was perfused with standard artificial cerebrospinal fluid. Time calibration: 5 ms between beginning of the traces and stimulus artefact. The vertical bar is 1 mV for the potentials (negativity down), and 100 mV/mm² for the CSD (the current sinks are down, indicated in black)

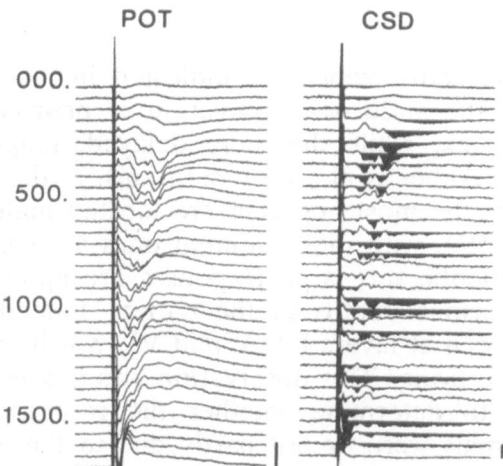

Fig. 4. Laminar potential (POT) and current source density (CSD) profiles in response to single shock stimulations of the white matter in slices of sensory motor cortex of an epileptic rat. The recording electrode was displaced by steps of 50 µm. the slice was perfused with standard artificial cerebrospinal fluid. Time calibration: 5 ms between beginning of the traces and stimulus artefact. The vertical bar is 2 mV for the potentials (negativity down), and 100 mV/mm² for the CSD (the current sinks are down, indicated in black). The synaptic responses are longer and much more extended, especially in the middle and deep layers, than in control animals

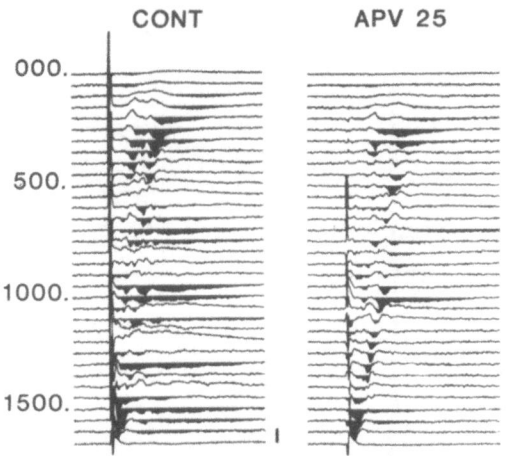

Fig. 5. Current source density profiles in response to single shock stimulations of the white matter in a slice of sensory motor cortex of an epileptic rat. The recording electrode was displaced by steps of 50 μm. the slice was perfused with standard artificial cerebrospinal fluid. CONT: the slice was perfused with standard ACSF. APV 25: the same slice was perfused with the same fluid, this time containing in addition some 2-amino-phosphonovalerate, a specific antagonist of glutamate at the N-methyl-D-aspartate receptor, at a concentration of 25 μM. Time calibration: 5 ms between beginning of the traces and stimulus artefact. The vertical bars are 100 mV/mm² (the current sinks are down, indicated in black). The synaptic responses are reduced in duration, mainly at the middle and deep layers

correspond to the active zones are indicated in black. The earlier short sinks are due to the antidromic activation of neurons located at various depths in the cortex, while the more delayed, longer sinks are due to synaptic currents (Abbès et al., in preparation). In the cortex of the control rat, the responses the most apparent are located mainly in the superficial layers (layers II–III), with more discreet ones at layers IV and V. In the cortex of the epileptic rat, large responses are apparent not only at the surface but at various depths in the cortex. Indeed, responses of long duration can be seen at layers IV, V and VI, which can certainly promote onset of repetitive firing. This observation was a constant finding in many slices. In order to determine whether the responses thus observed in the epileptic animals were or not partly due to the activation of NMDA receptors, we repeated the same protocol on the same slice, but after having added in the bath a selective antagonist of NMDA, the 2-amino-phosphonovalerate (APV). The result is shown in Fig. 5: a large part of the synaptic responses, especially at the middle and deep layers are largely reduced in the presence of 25 μM APV. It is interesting to note that the reduction of the response duration is more marked in the deep layers than in the superficial ones. Similar experiments performed on slices taken from control animals show that the NMDA-dependent component of the synaptic responses is small in these animals. These findings show that the NMDA-dependent components of the synaptic responses are more prominent and more widely distributed in the cortex of epileptic rats than that of control rats. The presence of such components gives rise to synaptic responses of longer duration at several cortical layers, thereby probably promoting repetitive activity.

Discussion

The main finding of this study is that NMDA responses are much more widely distributed in the cerebral cortex of rats with absence-like seizures than in the cortex of control rats. This finding agrees well with studies performed on the hippocampus of kindled rats, in which an increased responsiveness to NMDA was observed (Mody et al., 1988). It could also explain why, during generalized seizures in the photosensitive baboon Papio papio, the $[Ca^{2+}]_0$ falls to very low level (Pumain et al., 1985), due to the high calcium permeability of the NMDA-associated ionic channel.

The increase in sensitivity appears to be relatively selective for NMDA since a similar feature was not observed for the other agonists of glutamate, Quis or Ka, which are active on different receptors.

Concerning the responses to the ionophoretic applications of NMDA, the responses were the largest at the superficial layers and diminished monotonously towards the white matter (Fig. 1). This result is at variance with the result obtained in the cortex of normal Srague-Dawley rats: there, the largest responses were also observed at the superficial layers, but a second peak was apparent at a depth of about 1,200 µm (Pumain et al., 1987). We do not know yet to what is due such a difference, if not to strain difference.

Concerning the CSD study, the increase of the NMDA-dependent component of the synaptic responses can be interpreted in terms of a reduced efficacy of inhibitory mechanisms: then the duration of excitatory responses should be increased and the expression of NMDA responses enhanced. However, the cortical layers at which the augmented synaptic responses were observed correspond well with those displaying an increased sensitivity to direct applications of NMDA, namely the middle and deep layers. In addition, although we did not perform detailed studies of inhibition in the cortex of the epileptic rats, the inhibitory mechanisms did not appear to be profoundly altered. Nevertheless, this possibility cannot be entirely discarded.

In the strain of epileptic rats used in the present study, spike-and-waves discharges occur in the neocortex and lateral thalamus, but not in the limbic system (Vergnes et al., 1982). The question then arises to decide whether it is the cortex or the thalamus which is primarily involved in the triggering of the epileptic seizures. We cannot yet solve this problem, but our results show clearly that there is a modification of the synaptic transmission in the neocortex, suggesting that it has the major role in the development of the epileptic process.

These findings show further that the NMDA-dependent components of the synaptic responses are more prominent and more widely distributed in the cortex of epileptic rats than that of control rats. The presence of such components gives rise to synaptic responses of longer duration at several cortical layers, thereby probably promoting repetitive activity. In the epileptic Strasbourg rats, the enhancement of NMDA responses in middle and deep layers is quite reminiscent of a similar enhancement observed in the cortex of rats bearing a chronic cobalt focus (Pumain et al., 1986). This came as a surprise since these two epilepsies are quite different. However, a similar finding was also observed in epileptogenic cortical zones in man, and it may be suggested that a common mechanism may underlie chronic epileptogenesis, both in focal epilepsies, arising from zones with demonstrable lesions and in non-lesional, generalized epilepsies.

Acknowledgement

This work was supported by a grant from INSERM (CAR n° 4900019).

References

Abbes S, Louvel J, Lamarche M, Pumain R (1991) Laminar analysis of the origin of the various components of evoked potentials in slices of rat sensory motor cortex. Electroencephalogr Clin Neurophysiol 80: 310–320

Ammann D (1985) Ion-selective microelectrodes. Springer, Berlin Heidelberg New York Tokyo, p 346

Anderson WW, Lewis DV, Swartzwelder HS, Wilson WA (1986) Magnesium-free medium activates seizure-like events in the rat hippocampal slice. Brain Res 398: 215–219

Ascher P, Nowak L (1988a) Quisqualate- and kainate-activated channels in mouse central neurones in culture. J Physiol 399: 227–245

Ascher P, Nowak L (1988b) The role of divalent cations in the N-methyl-D-aspartate responses of mouse central neurones in culture. J Physiol 399: 247–266

Avoli M, Louvel J, Pumain R, Olivier A (1987) Seizure-like discharges induced by lowering $[Mg^{2+}]_0$ in the human epileptogenic neocortex maintained in vitro. Brain Res 417: 199–203

Avoli M, Gloor P, Kostopoulos (1990) Focal and generalized epileptiform activity in the cortex: in search of differences in synaptic mechanisms, ionic movements, and long-lasting changes in neuronal excitability. In: Avoli M, Gloor P, Kostopoulos G, Naquet R (eds) Generalized epilepsy. Neurobiological approaches. Birkhäuser, Boston Basel Berlin, pp 238–253

Croucher MJ, Collins JF, Meldrum BS (1982) Anticonvulsant action of excitatory amino acid antagonists. Science 216: 899–901

Freeman JA, Nicholson C (1975) Experimental optimization of current-source density technique for anuran cerebellum. J Neurophysiol 38: 369–382

Heinemann U, Konnerth A, Pumain R, Wadman WJ (1986) Extracellular calcium and potassium concentration changes in chronic epileptic brain tissue. Adv Neurol 44: 641–661

Louvel J, Pumain R, Roux FX, Chodkievicz JP (1992) Recent advances in understanding epileptogenesis in animal models and in humans. Adv Neurol 57: 517–524

Lux HD, Neher E (1973) The equilibration time course of $[K^+]_0$ in cat cortex. Exp Brain Res 17: 190–205

MacDermott AB, Mayer ML, Westbrook GL, Smith SJ, Barker JL (1986) NMDA-receptor activation increases cytoplasmic calcium concentration in cultured spinal cord neurones. Nature 321: 519–522

Marescaux C, Micheletti G, Vergnes M, Depaulis A, Rumbach L, Warter JM (1984) A model of chronic spontaneous petit mal-like seizures in the rat: comparison with pentylenetetrazol-induced seizures. Epilepsia 25: 326–331

Marescaux C, Vergnes M, Micheletti G (1984) Antiepileptic drug evaluation in a new animal model: spontaneous petit mal epilepsy in the rat. Fed Proc 43280: 280–281

Mayer ML, Westbrook GL (1987) The physiology of excitatory amino acids in the vertebrate central nervous system. Prog Neurobiol 28: 197–276

Mayer ML, Westbrook GL (1988) Permeation and block of N-methyl-D-aspartic acid receptor channels by divalent cations in mouse cultured central neurones. J Physiol 394: 501–527

Mayer ML, Westbrook GL, Guthrie PB (1984) Voltage-dependent block by Mg^{2+} of NMDA responses in spinal cord neurones. Nature 309: 261–263

Meldrum BS, Croucher MJ, Cuczwar SJ, et al (1983) A comparison of the anticonvulsivant potency of +2-amino-5-phosphonopentanoic acid and +2-amino-7-phosphonoheptanoic acid. Neuroscience 9: 925–930

Mitzdorf U (1985) Current source-density method and application in cat cerebral cortex: investigation of evoked potentials and EEG phenomena. Physiol Rev 65: 37–100

Mitzdorf U, Singer W (1978) Prominent excitatory pathways in the cat visual cortex (A 17 and A 18): a current source density analysis of electrically evoked potentials. Exp Brain Res 33: 371–394

Mitzdorf U, Singer W (1979) Excitatory synaptic ensemble properties in the visual cortex of the macaque monkey: a current source density analysis of electrically evoked potentials. J Comp Neurol 187: 71–84

Mody I, Stanton PK, Heinemann U (1988) Activation of N-methyl-D-aspartate receptors parallels changes in cellular and synaptic properties of dentate gyrus granule cells after kindling. J Neurophysiol 59: 1033–1054

Nicholson C, Freeman JA (1975) Theory of current-source density analysis and determination of conductivity sensor for anuran cerebellum. J Neurophysiol 38: 356–368

Nowak L, Bregestovski P, Ascher P, Herbet A, Prochiantz A (1984) Magnesium gates glutamate-activated channels in mouse central neurons. Nature 307: 462–465

Pumain R, Heinemann U (1985) Stimulus- and amino-acid induced calcium and potassium changes in rat neocortex. J Neurophysiol 53: 1–16

Pumain R, Menini C, Heinemann U, Louvel J, Silva-Barrat C (1985) Chemical synaptic transmission is not necessary for epileptic seizures to persist in the baboon Papio papio. Exp Neurol 89: 250–258

Pumain R, Louvel J, Kurcewicz I (1986) Ionic concomitants in chronic epilepsies. In: Speckmann E-J, Schulze H, Walden J (eds) Epilepsy and calcium. Urban and Schwarzenberg, München Wien Baltimore, pp 207–225

Pumain R, Kurcewicz I, Louvel J (1987) Ionic changes induced by excitatory amino acids in the rat cerebral cortex. Can J Physiol Pharmacol 65: 1067–1077

Richardson TL, Turner RW, Miller JJ (1987) Action potential discharge in hippocampal CA1 pyramidal neurons: current source-density analysis. J Neurophysiol 58: 981–996

Stanton PK, Jones RSG, Mody I, Heinemann U (1987) Epileptiform activity induced by lowering extracellular $[Mg^{2+}]$ in combined hippocampal-enthorhinal cortex slices: modulation by receptors for norepinephrine and N-methyl-D-aspartate. Epilepsy Res 1:53–62

Vergnes M, Marescaux C, Micheletti G, et al (1982) Spontaneous paroxysmal electroclinical patterns in rat: a model of generalized non-convulsive epilepsy. Neurosci Lett 33: 97–101

Vergnes M, Marescaux C, Depaulis A, Micheletti G, Warter JM (1987) Spontaneous spike and wave discharges in thalamus and cortex in a rat model of genetic petit mal-like seizures. Exp Neurol 96: 127–136

Vergnes M, Marescaux C, Depaulis A, Micheletti G, Warter JM (1990) Spontaneous spike-and-wave discharges in Wistar rats: a model of genetic generalized nonconvulsive epilepsy. In: Avoli M, Gloor P, Kostopoulos G, Naquet R (eds) Generalized epilepsy. Neurobiological approaches. Birkhäuser, Boston Basel Berlin, pp 238–253

Walther, H, Lambert JDC, Jones RSG, Heinemann U, Hamon B (1986) Epileptiform activity in combined slices of the hippocampus, subiculum and entorhinal cortex during perfusion with low magnesium medium. Neurosci Lett 69: 156–161

Authors' address: Dr. R. Pumain, Unité de Recherches sur l'Epilepsie, INSERM U 97, 2ter, rue d'Alésia, F-75014 Paris, France

J Neural Transm (1992) [Suppl] 35: 109–124

Possible mechanisms underlying hyperexcitability in the epileptic mutant mouse tottering

G. K. Kostopoulos and **C. T. Psarropoulou**

Department of Physiology, University of Patras Medical School, Patras, Greece

Summary. Tottering mice present a useful experimental model of genetically determined generalized epilepsy of the absence type. In electrophysiological recordings from hippocampal slices in vitro we found that the postsynaptic excitability (firing threshold) of pyramidal neurons in the CA_1 area of tg/tg slices was significantly higher than that of normal slices. In spite of this hyperexcitability, in vitro epileptiform discharges were not observed spontaneously, or upon provocation by intracellular depolarizing pulses, or in response to moderate elevations (+2 mM) in extracellular potassium. The latter elevations actually induced significantly smaller increases in the CA_1 synaptic responses of tg/tg as compared to normal slices. The hyperexcitability of tottering neurons could not be explained in terms of altered membrane electrical properties or any reduction of synaptic inhibition or increased capacity for long-term potentiation. Responses to noradrenaline, histamine and adenosine, as well as to the release of N-methyl-D-asparate channels — by eliminating Mg^{2+} — were comparable in tg/tg and normal slices. These studies show that hyperexcitability can be co-inherited with epilepsy and in this model its expression can be maintained in vitro. The neuronal mechanism of this expression remains elusive, as it does not appear to include some features known to be shared by experimental models of chemically or electrically induced epilepsy.

Introduction

Epilepsy was already proposed by Hughlings Jackson to be "a hyperphysiological state of the brain" (Jackson, 1931). Subsequent developments in electrophysiological research defined this state further as neuronal hyperexcitability. The exact nature of the association between the excitability of central neurons and epilepsy, as well as the underlying mechanisms, remains elusive. Documentation of hyperexcitability has been based mainly on the exaggerated membrane potentials underlying epileptiform discharges recorded during experimentally induced seizures. The balance of evidence from decades of research in epileptology presented disinhibition as the most probable mechanism underlying epileptogenic hyperexcitability (Dingledine and Gjerstad, 1980; Krnjevic, 1983; Lopes da Silva, 1987; Spencer and Kandel, 1969; Swartzkroin and Prince, 1980). It is conceivable, however, that this proposal is partly biased by the experimental conditions experimenters create in order to induce epilepsy (Malouf et al., 1990), which may differ substantially from those underlying naturally occurring epilepsy. Indeed, hyperexcitability has still to be unequivocally demonstrated in a model of naturally occurring epilepsy, i.e. independently of any convulsant experimental provocation. Furthermore, there is now evidence that different types of epilepsy may develop through entirely different mechanisms (Gloor and Fariello, 1988; Avoli et al., 1990: article by Avoli et al.). Each experimental model should therefore be examined separately and under the light of the particular type of epilepsy it models.

Epilepsy is a multifactorial disease (Gloor et al., 1982). The genetic predisposition, one of these factors, is often expressed by the presence or provocation of spike-and-wave discharges (SW), an EEG feature proposed to characterize the first degree of epileptogenesis. The tottering mouse offers a genetic model of generalized epilepsy of the absence type with SW (Noebels and Sidman, 1979; Kaplan et al., 1979; Noebels, 1984; see review by Kostopoulos, this volume). The recent demonstration of gene-linked neuronal hyperexcitability in this mutant, which was maintained in vitro (Kostopoulos and Psarropoulou, 1990), raised some hopes of elucidating the neuronal mechanisms underlying hyperexcitability and its role in epileptogenesis. The experiments to be described here are part of the efforts of our laboratory towards this goal. They were guided by (a) a previous experience in other animal models of a similar general type of epilepsy (Kostopoulos et al., 1981; Gloor et al., 1990; Avoli et al., 1990) and (b) by the seminal finding of Levitt and Noebels (1981) that tottering mice show an abnormal proliferation of noradrenergic terminals — this abnormality being necessary for the expression of the disease (Noebels, 1984), as well as by more recent findings concerning the fascinating cellular phenotype of this mutant (see ref. in Kostopoulos, this volume).

Material and methods

A colony of tottering mice was raised locally from breeding pairs obtained from Jackson Laboratories (Maine, USA). Epileptic mice (tg/tg) were identified both by their abnormal gait and by the incidence of focal motor seizures (Noebels and Sidman, 1979). Electrographic studies in our laboratory (Kostopoulos et al., 1987) convinced us that these mice would invariably present EEG and behavioural signs of absence epilepsy.

We used conventional biochemical (Psarropoulou et al., 1987; Angelatou et al., 1990), as well as electrophysiological techniques for in vitro recordings from hippocampal slices conducted intracellularly (Kostopoulos et al., 1988) and extracellularly (Kostopoulos and Psarropoulou, 1990). Details of material and methods used are explained in these publications.

Results

1. Intracellular studies

We have reported a comparative study of the spontaneous oscillations of the membrane potential and other basic electrophysiological properties of the membrane of hippocampal CA_1 pyramidal neurons in slices from tg/tg and normal mice (Kostopoulos et al., 1988). No paroxysmal depolarization shifts or epileptiform discharges, either spontaneous or in response to electrical stimulation, were noted in any of the neurons.

Experimentally induced noradrenergic hyperinnervation of trigeminal motoneurons (Vornov and Sutin, 1986) significantly alters the electrical properties of their membrane. They were about 3 mV more hyperpolarized with reduced membrane input impedance. Evoked excitatory postsynaptic potentials (EPSP) were potentiated, while their rise time was increased in consistency with the morphological demonstration of replacement by the proliferated noradrenergic terminals of afferent input from its synaptic sites normally proximal to the soma, to more distal dendritic locations (Hemmendinger and Moore, 1983). One might expect to observe similar findings in hippocampal neurons of the tottering mutant, since they are also hyperinnervated by noradrenergic terminals. Also, kindling epilepsy has been reported to induce an increase of neuronal membrane input impedance, which is maintained in vitro (Mody et al., 1988). The resting membrane potential, input impedance and time constant, however, were all found normal in CA_1 pyramidal neurons of tg/tg hippocampal slices (Kostopoulos et al., 1988). The decay constant of the after-hyperpolarization following an intracellular depolarizing pulse was not significantly different in tg/tg vs. normal slices.

These results make it unlikely that the tg mutation is associated to a defect in resting ion permeabilities, or to after-hyperpolarizing currents, which are commonly observed to limit prolonged and excessive depolarizations of the neuronal membrane (Kostopoulos et al., 1988).

2. Field potential studies

The negative findings emerging from the intracellular recordings suggested that if the tg mutation commanded any epileptogenic hyperexcitability in hippocampal neurons, this either might be very subtle to be found in a study of single neurons, or was expressed only under conditions of synchronous synaptic activation of large populations of neurons, or both.

2.1 Excitability curves

In order to study the excitability of large populations of CA_1 pyramidal neurons, we constructed input/output (I/O) curves from field synaptic

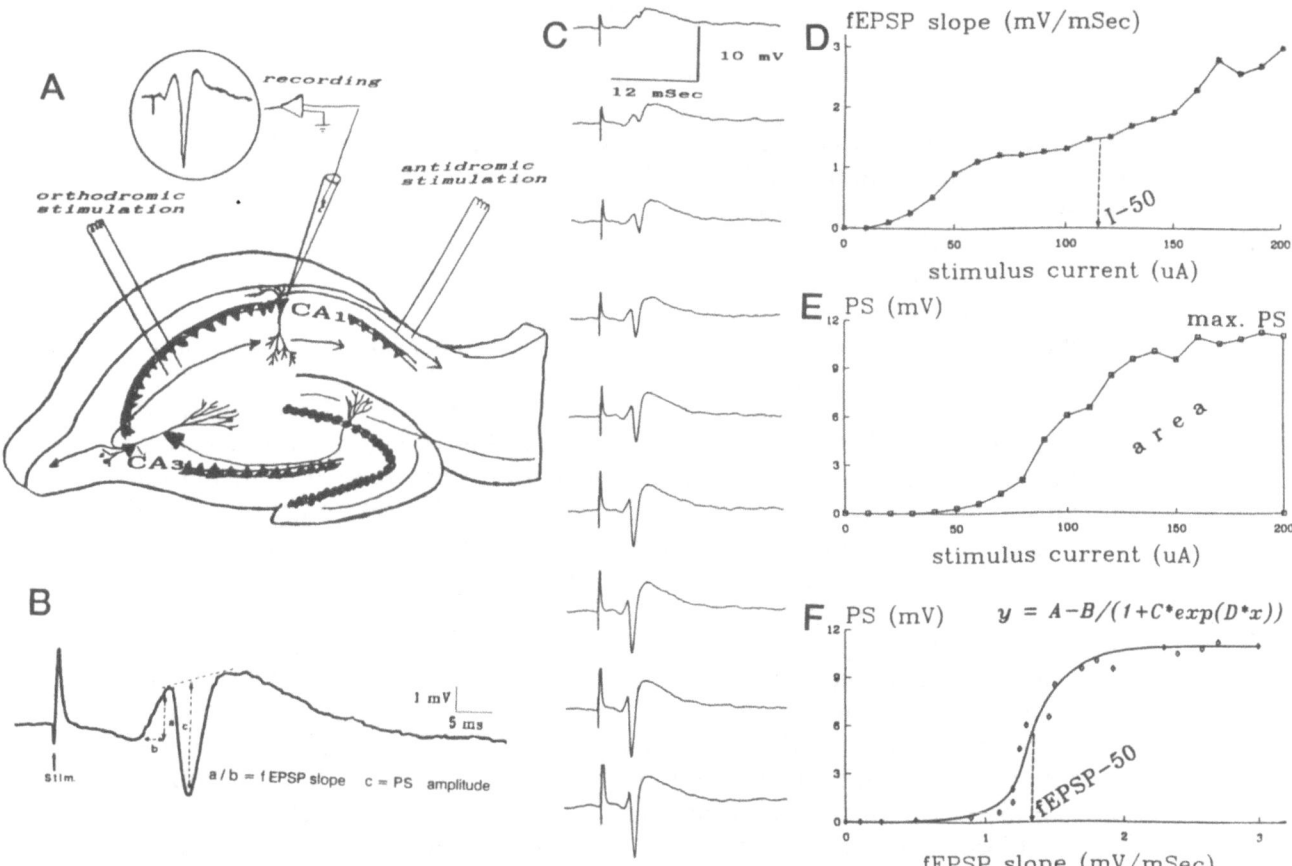

Fig. 1. Experimental arrangements for recording field potentials from the CA_1 area of hippocampal slices and methods used to quantify neuronal excitability. **A** Simulating electrodes in stratum radiatum and alveus evoke field responses in the pyramidal layer. **B** Measurement of fEPSP and PS. **C** Some of the responses to increasing stimulus strength — top to bottom — used for constructing the input/output curves in D–F. **D** Plot of the fEPSP values against the stimulus intensity (I) used to evoke them in a series of stimuli of increasing strength. Some of these responses are shown in C. The I_{50} is calculated as the stimulus intensity needed to produce a half-maximal fEPSP response. **E** The PS values of the responses to the same set of stimuli are similarly plotted. The maximum PS and the area of the polygon is measured as indicated. **F** The PS amplitude of each response to the same set of stimuli as in D and E is plotted against the corresponding fEPSP value. A computer program uses the equation shown to fit a sigmoid line on these points and to calculate the curve's maximal rising slope as well as the projection of the point of maximal slope onto the x-axis ($fEPSP_{50}$)

responses recorded in their layer (Fig. 1A). Stimuli of successively increasing and decreasing strength were delivered in stratum radiatum (constant current, $I = 0$–0.3 mA, 0.2 mS, Fig. 1C). The rising slope of the field excitatory postsynaptic potential (fEPSP) and the amplitude of the population spike (PS) were measured (Fig. 1B). From the polygons fEPSP/I (Fig. 1D) and PS/I (Fig. 1E), we calculated the values of maximal fEPSP

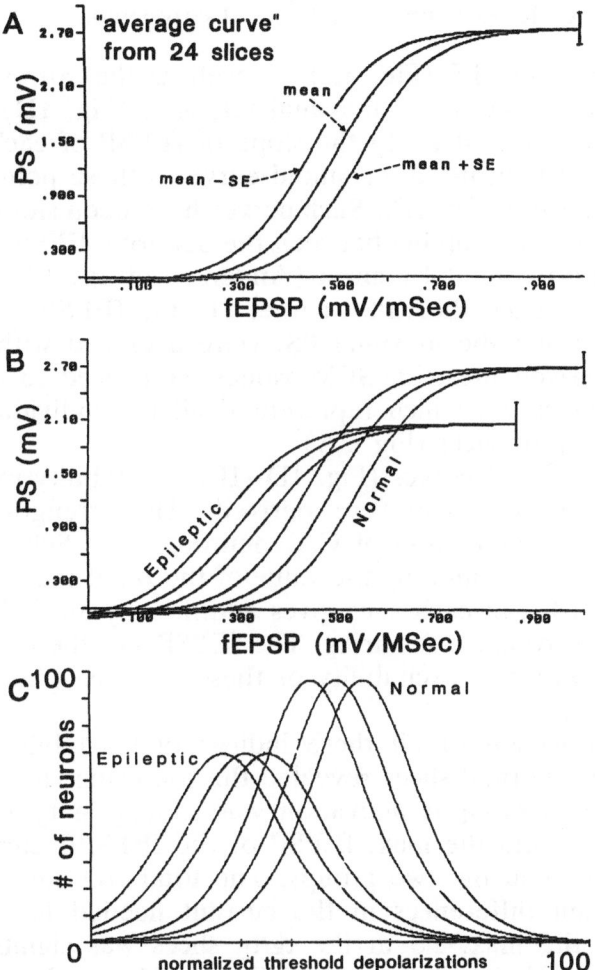

Fig. 2. Hippocampal CA_1 pyramidal neurons of epileptic tg/tg mice have higher in vitro postsynaptic excitability compared to normal, as demonstrated by the leftward shift of their PS/fEPSP curves. **A** After a sigmoid line was fitted to the experimental data from I/O curves in 24 individual normal slices, as described in Fig. 2, the "average postsynaptic excitability curve" representing data from these slices was constructed by averaging the corresponding parameters from the curves of each slice. The two parameters (A and B), which determine the y-axis expansion of each curve, as well as its max. slope and its $fEPSP_{50}$, were averaged. From the means of the latter two values we determined the "average C" and "average D". A triplet of such curves is constructed using the Mean ± SEM of the $fEPSP_{50}$. The vertical bar indicates ± 1 SEM of the max. PS. In a similar way we constructed the average postsynaptic excitability curve of the group of 29 tg/tg slices. The latter was found to be shifted down and to the left of that of normal slices. Both displacements were statistically significant. **B** Graphical representation of the assumptions underlying the difference between I/O sigmoid curves as explained in text. Differentiation of the curves shown in A illustrates the probability density of spike thresholds in neurons of the two groups studied. On the x-axis the values of the fEPSP recorded in an I/O experiment actually reflect the normal distribution of the corresponding (Andersen et al., 1971) increase in amplitude depolarizations, which individual neurons need to undergo before firing action potentials. The depolarization step corresponding to the experimentally found mean $fEPSP_{50}$ (0.49 mV/mS) is given a value of 100% for normalization purposes. On the y-axis the number of neurons firing at each depolarization step is similarly expressed relatively to the number of neurons in the group of normal slices firing at the "mean threshold depolarization". A lower mean spike threshold in tg/tg slices as compared to normals as well as a smaller maximal number of neurons available for excitation is indicated

(Fig. 1D), and maximal PS (Fig. 1E), as well as the corresponding I_{50}s (current needed to evoke a half-maximal response (Fig. 1D). Finally, we correlated the amplitude of PS to the slope of fEPSP of each trace in an I/O experiment and then fitted a sigmoid curve to these points (according to the equation shown in Fig. 1F). Such curves have been shown to actually represent the functional coupling between the dendritic EPSP and the spike trigger zone of the pyramidal neuron (Andersen et al., 1980). The three values that describe each PS/fEPSP curve, i.e. the $fEPSP_{50}$, the maximal slope of the curve and the maximal PS, were averaged within groups of slices, and the derived mean $\pm$ SEM values were used to construct the "average" sigmoid curve, which represented all the individual PS/fEPSP curves in each group of slices (Fig. 2A).

From each of the I/O curves (Fig. 1D–1F), excitability indices of particular functional significance were derived. The strength of synaptic activation of a large population of CA_1 neurons upon Schaffer collateral stimulation is best represented by the value of the I_{50} of the fEPSP/I curve. The max. PS recorded in each slice gives additionally an indication of the number of neurons available for firing. The $fEPSP_{50}$ of the PS/fEPSP curve measures the postsynaptic excitability of these neurons (Andersen et al., 1971, 1980).

Statistical comparison of all these indices of excitability in the two groups of tg/tg and normal slices revealed the following (Kostopoulos and Psarropoulou, 1990): synaptic activation was similar in tg/tg and normal mice, since the I_{50}s and the max. fEPSP of the fEPSP/I curves were not significantly different in the two groups. The total I/O curves (PS/I) also showed insignificant differences in the current needed for half-maximal firing. However, the max. PS in the tg/tg slices was significantly lower (tg/tg: 1.97 ± 0.25 mV, n = 29; normal: 2.72 ± 0.22, n = 24; p = 0.0044 by the Mann-Whitney test).

The important difference apparently resided in the postsynaptic excitability of CA_1 neurons, which was significantly higher in the slices from epileptic mice. This is shown by the smaller fEPSP needed to evoke the same amplitude of PS ($fEPSP_{50}$ in the fEPSP/PS curves was in tg/tg: 0.30 ± 0.01, n = 29; in normal 0.49 ± 0.05, n = 24; p = 0.0024). The difference in postsynaptic excitability between tg/tg and normal slices is shown graphically in Fig. 2B, where the average postsynaptic excitability curve for each group is calculated. Each triplet shows the respective mean $\pm$ SEM of the curve, while the vertical bars show the SEM of the max. PSs.

Neuronal hyperexcitability was thus documented in an experimental model of naturally occurring epilepsy. The significant leftward shift of the average PS/fEPSP curve in tg/tg slices as compared to the normal ones indicated that most CA_1 pyramidal neurons in these slices fired in response to relatively smaller synaptic depolarizations. The smaller max. PS in tg/tg slices can be tentatively interpreted as being due to a smaller number of neurons available for firing.

The assumptions on which we based our interpretation of the data (Andersen et al., 1971, 1980) are illustrated by the hypothetical curve of Fig. 2C. It is thus suggested that the mechanisms of epileptogenic hyperexcitability in tg/tg mice should primarily be sought among those mechanisms expressed at the postsynaptic level. We examined the most apparent of those putative mechanisms, as described below.

2.2 Study of recurrent inhibition

PS amplitude is to a large extent controlled by the disynaptic IPSP that follows activation of the same afferents as that causing the monosynaptic EPSP. Noebels and Rutecki (1990, in press) reported that hippocampal neurons of tg/tg mice show relatively prolonged epileptiform discharges, when exposed to epileptogenic agents. The latter characteristic of tottering hippocampus was mimicked only by picrotoxin, a GABA antagonist among other convulsants, thus suggesting a functional impairment of $GABA_A$-mediated inhibition. GABAergic inhibition could be reduced in tottering brain by an action of the excessively released noradrenaline (Madison and Micoll, 1988; Levitt and Noebels, 1981).

We therefore examined whether the epileptogenic hyperexcitability in the tottering mutant is secondary to or in any way associated with a change in inhibition of neuronal activity. Paired-pulse inhibition, which has been shown to be altered in human epileptogenic hippocampus (Kahn et al., 1989) and in rats following kindling (Lopes da Silva, 1987; Tuff et al., 1983) and other types of experimentally induced epilepsy (Kapur and Lothman, 1989), was examined. The rationale for using this protocol is that the decrease in amplitude of the second of a pair of evoked responses at short inter-pulse intervals is attributed to the activation of feed-forward and recurrent inhibitory circuits (Buszaki, 1984; Newberry and Nicoll, 1984; Andersen et al., 1964).

In vitro electrophysiological experiments in hippocampal slices examined several aspects of neuronal inhibition in this mutant (Kostopoulos and Antoniadis, in preparation), using both the antidromic-orthodromic and the double-orthodromic paired-pulse stimulation protocols in the CA_1 area. Antidromic inhibition (Fig. 3) was quite strong at 5- and 10-msec intervals (70%) and lasted up to at least 320 msec (15%). It also caused a shift to the right of the I/O curves (mean % increase in I_{50} was 88.3 at 10-msec inter-pulse interval). Paired-pulse stimulation with two orthodromic stimuli produced a biphasic response: very early inhibition (60% at 2.5 msec) and late excitation (about 40% at 20–160 msec). Statistical comparison of these indices of phasic inhibition suggested that the inhibitory control of excitability is as strong in tg/tg slices as in normals. A tendency towards a longer duration of antidromic inhibition in the tg/tg slices was actually observed (Fig. 3).

116   G. K. Kostopoulos and C. T. Psarropoulou

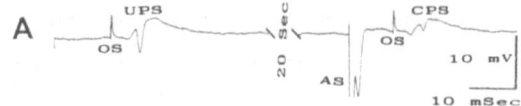

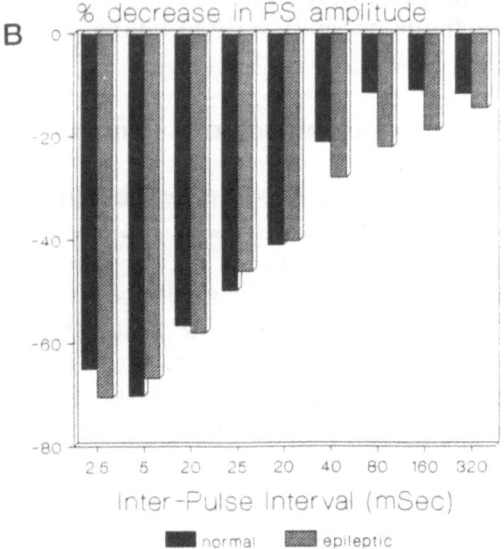

Fig. 3. Similarity of recurrent inhibition in normal and tg/tg hippocampal slices. **A** Field responses in the CA$_1$ pyramidal layer to stimuli delivered at locations shown in Fig. 2A. OS and AS, artefacts from ortho- and antidromic stimuli. UPS and CPS, unconditioned population spike and PS conditioned by the preceding antidromic response. UPS and CPS elicited in an alternating fashion every 20 s. **B** Mean decrease of PS amplitude [(UPS − CPS) × 100/UPS] due to the recurrent inhibition evoked by the preceding antidromic response, as a function of the interval between antidromic and orthodromic stimuli

It has been further proposed that the development of epilepsy may not depend so much on the integrity of inhibitory circuits as on the degree of activity-dependent "fatigue" of this inhibition (Ben-Ari et al., 1979; Kapur and Lothman, 1989; McCarren and Alger, 1985). We have therefore compared this phenomenon in slices from normal and from epileptic mice. Immediately following a train of antidromic stimuli (8 sec, 5 Hz) a 42% decrease in antidromic inhibition was observed, inhibition returning to normal within about a minute. No significant differences between hippocampal slices taken from epileptic or normal mice were observed in this respect either. However, the train of antidromic pulses causes a significant and overlasting decrease in the unconditioned orthodromic PS amplitude only in the tg/tg slices, suggesting a stronger potentiation of tonically active inhibitory interneurons in these mice.

It is concluded that the mechanisms underlying epileptogenic hyperexcitability in the tottering mutant mouse may not include a failure of phasic inhibition, at least in the CA_1 area of hippocampus, while there are signs of a more easily activated long-lasting or tonic inhibitory mechanism.

2.3 Study of LTP

Long-term potentiation (LTP) (Bliss and Lynch, 1988) involves an increased neuronal excitability not dependent on any decrease in synaptic inhibition (Haas and Rose, 1985). Since mechanisms underlying LTP have been linked to the two conditions characterizing the tottering mutation, namely epilepsy (Cain, 1989) and increased noradrenergic innervation (Hopkins and Johnston, 1988), we decided to examine the possibility that tottering mice have an increased capacity for LTP.

LTP was produced in two complementary ways, with regard to the neuronal mechanisms supposedly involved. Tetanic stimulation produces LTP through a complex succession of postsynaptic and presynaptic mechanisms (Collingridge and Bliss, 1987; Bliss, 1990), while elimination of Mg^{2+} ions from the perfusion medium appears to involve primarily post-synaptic elements (Psarropoulou and Kostopoulos, 1990). Comparison of the LTP present 10 minutes after tetanic stimulation (100 Hz, 500 ms) of stratum radiatum in the CA_1 area of hippocampal slices taken from tg/tg and normal mice revealed no significant differences (Kostopoulos et al., 1988). Two hours after a 30-minute perfusion with Mg^{2+} — free medium, the PS response of the CA_1 neurons to stratum radiatum stimulation was significantly increased in both normal and tg/tg slices. As the field EPSP did not show any long-term change, the $fEPSP_{50}$ of the PS/fEPSP curves, which is an index of postsynaptic excitability, was increased (by mean $\pm$ SD = 25 $\pm$ 16% in 12/12 normal slices and by 30 $\pm$ 7% in 13/13 tg/tg slices, the difference being non-significant).

2.4 Sensitivity to neurotransmitters and modulators

The demonstrated postsynaptic hyperexcitability could have resulted from a genetically determined change in the sensitivity of tg/tg neurons to certain neurotransmitter substances known to be involved in epileptogenesis.

A defect in GABAergic transmission has been implicated in several epileptic models (Krnjevic, 1983; Lopez da Silva, 1987). Comparing brains of tg/tg and normal mice, however, we found no statistical difference in the binding of $GABA_A$, $GABA_B$ and benzodiazepine receptors (Psarropoulou et al., 1987). Adenosine has been considered as an endogenous anti-epileptic substance (Dragunow, 1988; Kostopoulos, 1988). Although we found evidence that A_1 adenosine receptors are up-regulated after pentylene-tetrazol convulsions, no difference in the same receptors was found between tg/tg and normal brains (Angelatou et al., 1990). The in vitro electrophysiological response of tg/tg hippocampal neurons to adenosine (10–50 µM, Psarropoulou and Kostopoulos, in preparation) and to histamine (Kostopoulos et al., 1988) was quite comparable to that from hippocampi of normal mice. With regard to the integrity of the cholinergic system in tottering brain, we could measure normal choline acetyl-transferase activity in the cortex of these mice (Psarropoulou et al., 1987).

The proliferation of noradrenergic terminals in tottering brain could conceivably be an adaptive response to decreased sensitivity to noradrenaline. Alternatively, this increase in the number of terminals could be expected to down-regulate noradrenergic receptors (see Kostopoulos, this volume). However, on studying the in vitro intracellular and/or extracellular response of hippocampal CA_1 neurons to noradrenaline (both α_2 and β receptor agonists, Kostopoulos et al., 1988), we obtained no evidence of any significant difference between epileptic and normal mice. The demonstration of a normal sensitivity to noradrenaline in tg/tg slices contrasts with the lowered sensitivity to this transmitter displayed by pyriform slices taken from kindled rats (McIntyre and Wong, 1986). Furthermore, it provides functional confirmation of the morphological (Levitt and Noebels, 1981) and biochemical (Levitt et al., 1984) evidence favouring a truly increased noradrenergic neurotransmission in the tottering mutant.

Increased involvement of N-methyl-D-aspartate (NMDA) receptors in synaptic transmission has been directly implicated in epileptogenesis in several experimental models of epilepsy, including kindling (Dingledine et al., 1986; Meldrum et al., 1983; Mody et al., 1988). An indirect measure of the functional state of NMDA receptors in the hippocampus of tottering mice was achieved by comparing the enhancement of the in vitro synaptic response in the CA_1 area after eliminating Mg^{2+} ions from the extracellular

Table 1. Percentage changes in the mean values of excitability indices of CA1 pyramidal cells from normal and tg/tg mice, induced by a 2 mM increase in $[K^+]_o$[a,b]

	Syn. activation fEPSP/I		Input/output PS/I			Postsyn. excitability PS/fEPSP	
	I_{50}	Area	I_{50}	Area	Max Ps	Max slope	$fEPSP_{50}$
Normal	−9.41 ±2.69	+17.43 ±3.58	−22.83 ±3.48	+41.41 ±10.15	+12.08 ±4.34	+25.62 ±10.71	−29.54 ±5.63
n	22	23	24	24	24	24	24
	*c	*		*	*	*	*
tg/tg	−1.57 ±2.53	+4.52 ±4.21	−20.05 ±4.41	+16.73 ±4.81	+1.89 ±3.71	−2.84 ±4.98	−13.42 ±7.38
n	17	19	19	19	19	19	19

[a] Percentage changes are expressed as Mean + S.E.M.
[b] +% increase and −% decrease as compared to control values (100%).
[c] Statistics were made with unpaired Student's t-test, * P < 0.05

fluid in normal vs. tg/tg slices. All indices of excitability were changed to the same extent in the two groups of slices (Psarropoulou and Kostopoulos, 1990). It is thus suggested that, unlike kindling (Mody et al., 1988), the neuronal mechanisms underlying tottering epilepsy do not include any increase in the availability of NMDA receptors.

2.5 Sensitivity to moderate elevation (+2 mM) in extracellular K^+

One of the most fundamental determinants of neuronal excitability is the extracellular concentration of potassium ions (Somjen, 1979). Experimental (Poolos et al., 1987) or mutagenic (Ganetzky and Wu, 1985) dereases in transmembrane potassium currents have been shown to increase excitability and eventually to promote the generation and spread of epileptiform discharges (Traynellis and Dingledine, 1988; Tancredi and Avoli, 1987). It was therefore deemed necesary to examine the response of tg/tg neurons to elevations in extracellular $[K^+]$. Moderate elevations in $[K^+]_o$ were preferred (from 3 mM to 5 mM), in order to avoid (a) inducing epileptiform discharges and (b) decreasing the excitability of CA_1 afferents, as has been reported for higher $[K^+]_0$ (Tancredi and Avoli, 1987; Noebels and Rutecki, 1990; Poolos et al., 1987).

The changes in the mean values of the excitability indices were significantly less pronounced in the group of tg/tg slices, being statistically significant in only 4/7 indices, compared to 6/7 indices in the group of normal slices, an indication of decreased responsiveness in the former group (Table 1). The statistical comparison of the percentage changes in excitability indices between the two groups showed that these changes were significantly ($p < 0.05$) less pronounced in the tg/tg slices for 6/7 indices (Table 1).

Discussion

The principal finding in these studies is an increased postsynaptic excitability of hippocampal CA_1 neurons in an experimental model of naturally occurring epilepsy, the tottering mouse. The questions immediately arising are (a) the relationship of this hyperexcitability to epileptogenesis in this mutant and (b) the underlying molecular mechanism.

We were not surprised to find no paroxysmal depolarization shifts in the intracellular records and epileptiform discharges in the field recordings from slices derived from epileptic animals. Studies in human temporal cortex slices from electrographically verified epileptogenic tissue have shown that such electrophysiological manifestations of epilepsy are not readily appreciable in vitro, at least not in their classical form observed in vivo (Avoli et al., 1992; Prince and Connors, 1986). What our negative results might actually demonstrate is that postsynaptic hyperexcitability is not by itself a sufficient condition for epileptogenesis, at least in the in vitro conditions of this study.

Hippocampal slices from tottering mice are, however, capable of organizing synchronous epileptiform discharges, when activated by different convulsant chemicals. In particular the discharges evoked by elevating extracellular potassium ion concentrations ($+10$ mM) or by 4-aminopyridine are of significantly longer duration in the CA_3 area of hippoampal tg/tg slices, as compared to slices taken from normal mice (Noebels and Rutecki, 1990). Interestingly, in our experiments non-epileptogenic ($+2$ mM) elevations in potassium under similar in vitro conditions cause a smaller increase in excitability in tg/tg CA_1 area, compared to normals. The difference between the two studies could be attributed to a gene-linked defect specific for the CA_3 and not the CA_1 area. Alternatively, the suspected defect in tg/tg slices (Noebels and Rutecki, 1990) might involve a potassium permeability, which is relatively unimportant under resting conditions, but can be activated after prolonged and excessive depolarizations, as can those underlying epileptiform discharges. It should be noted in this respect that the after-hyperpolarization following a strong 250-msec depolarization of CA_1 neurons was found to be of comparable duration in normal and tg/tg slices (Kostopoulos et al., 1988).

One cannot overemphasize the limitations imposed on the interpretation of our data, arising from the fact that they were obtained in vitro from a hippocampal circuit deprived of extrinsic inputs — most importantly of the influence of locus coeruleus terminals, which are proliferated in this mutant (Levitt and Noebels, 1981). However, the slice technique allowed precise control of the chemical micro-environment and the examination of synaptic activation separately from its impact on cell firing. The significant leftward shift of the average PS/fEPSP curve in the tg/tg slices, as compared to that of normals, suggested (Andersen et al., 1971, 1980) that the hyperexcitability of tg/tg neurons is primarily postsynaptic, i.e. that tg/tg neurons can fire in response to smaller depolarizations compared to normals. Several factors can possibly contribute to the latter: strictly intrinsic properties of the neuron, allowing better electrical coupling between a dendritic EPSP and the firing zone; control of firing by synaptic inhibition; altered response to general modulators of neuronal excitability; and an increased tendency to long-term potentiation. The mostly negative results we have obtained, while testing aspects of these factors, suggest that the observed in vitro hyperexcitability of tg/tg neurons may be due to such a novel arrangement of recurrent collaterals (Stanfield, 1989) that excitability and synaptic synchronization are enhanced without any change in the physiology of individual neurons.

Increased coupling between the synaptic and firing zones of a neuron can be achieved in many different ways (Jasper and Van Gelder, 1982). The present experiments render some of them unlikely to underlie the observed postsynaptic hyperexcitability. Resting membrane potential and other electrical characteristics of tg/tg CA_1 pyramidal neurons were found normal. Also normal was feedback and feed-forward inhibition in the CA_1 area. Thus, disinhibition, which has been shown to contribute to epileptogenesis in other experimental models (Dingledine and Gjerstad, 1980; Krnjevic, 1983; Lopes da Silva, 1987; Spencer and Kandel, 1966; Schwartzkroin and Prince, 1980), does not seem to be important for the development of this naturally occurring type of epilepsy. Responsiveness of NMDA receptors or noradrenergic and histamine receptors was also observed to be comparable in tg/tg and normal slices. Finally, it is interesting to note that adenosine receptors, which are significantly up-regulated in mice after pentylenetetrazole convulsions (Angelatou et al., 1990), are quite normal in tg/tg mice.

We would like to emphasize the important differences between our findings as described above and those from in vitro electrophysiological studies in slices from rats kindled to become epileptic. In such slices, CA_1 neurons have normal postsynaptic excitability and impaired synaptic inhibition (King et al., 1985; Lopez da Silva, 1987). Also, exposure to Mg^{2+} — free medium led to greater enhancement of the response in slices from kindled animals than normals (Mody et al., 1988), thus suggesting that kindling epilepsy is associated with an increase in the number of the availability of NMDA receptors. Tottering CA_1 neurons demonstrate increased postsynaptic excitability in the absence of decreased inhibition or increased "readiness" of NMDA receptor channels. Similar findings characterize cortical neurons in the feline model of generalized epilepsy with SW (Kostopoulos et al., 1981, 1983; Gloor et al., 1990). The present studies therefore support and extend the proposal that generalized epilepsy with SW is caused by distinct neuronal mechanisms in many respects different from those underlying partial epilepsy (Gloor and Fariello, 1988).

Acknowledgements

The studies described were supported by grants from the Greek Ministry of Research and Technology.

References

Angelatou F, Pagonopoulou O, Kostopoulos G (1990) Alterations of A1 adenosine receptors in different mouse brain areas after pentylentetrazol-induced seizures but not in the epileptic mutant mouse tottering. Brain Res 534: 251–256

Avoli M, Gloor P, Kostopoulos G, Naquet R (eds) (1990) Generalized epilepsy: neurobiological approaches. Birkhäuser, Boston

Avoli M, Hwa GGC, Drapeau G, Kostopoulos G, Perreault P, Olivier A, Villemeure J-G (1992) Electrophysiological analysis of human neocortex in vitro: experimental techniques and methodological approaches. Can J Neurol Sci (in press)

Ben-Ari Y, Krnjevic K, Reinhardt W (1979) Hippocampal seizures and failure of inhibition. Can J Physiol Pharmacol 57: 1462–1466

Bliss TVP (1990) Maintainance is presynaptic. Nature 346: 698–699

Bliss TVP, Lynch MA (1988) Long-term potentiation of synaptic transmission in the hippocampus: properties and mechanisms. In: Landfield PW, Deadwyller PW (eds) Long-term potentiation: from biophysics to behavior. Alan R Liss, New York, pp 3–72

Buzsaki G (1984) Feed-forward inhibition in the hippocampal formation. Prog Neurobiol 22: 131–153

Cain DP (1989) Long-term potentiation and kindling: how similar are the mechanisms? TINS 12(1): 6–10

Collingridge GL, Bliss TVP (1987) NMDA receptors — their role in long-term potentiation. Trends Neurosci 10: 288–293

Dingledine R, Gjerstad L (1980) Reduced inhibition during epileptiform activity in the in vitro hippocampal slice. J Physiol (Lond) 305: 297–313

Dingledine R, Hynes MA, King GL (1986) Involvement of N-methyl-D-aspartate receptors in epileptiform bursting in the rat hippocampal slice. J Physiol (Lond) 380: 175–189

Dragunow M (1988) Purinergic mechanisms in epilepsy. Prog Neurobiol 31: 85–108

Ganetsky B, Wu C-F (1985) Genes and membrane excitability in Drosophila. Trends Neurosci 8: 322–326

Gloor P, Fariello RG (1988) Generalized epilepsy: some of its cellular mechanisms differ from those of focal epilepsy. Trends Neurosci 11(2): 63–68

Gloor P, Metrakos J, Metrakos K, Andermann E, van Gelder N (1982) Neurophysiological, genetic and biochemical nature of the epileptic diathesis. Electroencephalogr Clin Neurophysiol [Suppl 35]: 45–56

Gloor P, Avoli M, Kostopoulos G (1990) Thalamocortical relationships in generalized epilepsy with bilaterally synchronous spike-and-wave discharge. In: Avoli M, Gloor P, Kostopoulos G, Naquet R (eds) Generalized epilepsy: neurobiological approaches. Birkhäuser, Boston, pp 190–212

Haas HL, Rose G (1984) The role of inhibitory mechanisms in hippocampal long-term potentiation. Neurosci Lett 47: 301–306

Hemmendinger LM, Moore RY (1983) Synaptic reorganization in rat motor trigeminal nucleus following neonatal 6-hydroxy-dopamine treatment. Soc Neurosci Abstr 9: 988

Hopkins WF, Johnston D (1988) Noradrenergic enhancement of long-term potentiation at mossy fiber synapses in the hippocampus. J Neurophysiol 59(2): 667–687

Jackson JH (1931) Selected writings of John Hughlings Jackson, vol 1. On epilepsy and epileptiform convulsions (edited by Taylor J). Hodder and Stroughton, London, p 500

Jasper HH, van Gelder NM (eds) (1983) Basic mechanisms of neuronal hyperexcitability. Alan R Liss, New York, pp 495

Jonzon B, Fredholm BB (1984) Adenosine receptor mediated inhibition of noradrenaline release from slices of the rat hippocampus. Life Sci 35: 1971–1979

Kahn SU, Wilson CL, Isokawa-Akesson M, Babb TL, Levesque MF (1989) Increased paired-pulse inhibition in the epileptogenic human temporal lobe. Soc Neurosci Abstr 15(1): 236

Kaplan BJ, Seyfred TN, Glaser GH (1979) Spontaneous polyspike discharges in an epileptic mutant mouse (tottering). Exp Neurol 66: 577–586

Kapur J, Lothman EW (1989) Loss of inhibition precedes delayed spontaneous seizures in the hippocampus after tetanic electrical stimulation. J Neurophysiol 61(2): 427–434

King GL, Dingledine R, Giachinno JL, McNamara JO (1985) Abnormal neuronal excitability in hippocampal slices from kindled rats. J Neurophysiol 54(5): 1295–1304

Kostopoulos G (1988) Adenosine: a molecule for synaptic homeostasis? In: Avoli M, Reader TA, Dykes RW, Gloor P (eds) Neurotransmitters and cortical function. Plenum Press, New York, pp 415–435

Kostopoulos G (1992) The tottering mouse: a critical review of its usefulness in the study of the neuronal mechanisms underlying epilepsy (this volume)

Kostopoulos G, Psarropoulou C (1990) Increased postsynaptic excitability in hippocampal slices from the tottering epileptic mutant mouse. Epilepsy Res 6: 49–55

Kostopoulos G, Gloor P, Pellegrini A, Gotman J (1981) A study of the transition from spindles to spike and wave discharge in feline generalized penicillin epilepsy: microphysiological features. Exp Neurol 73: 55–77

Kostopoulos G, Avoli M, Gloor P (1983) Participation of cortical recurrent inhibition in the genesis of spike and wave discharges in feline generalized penicillin epilepsy. Brain Res 227: 101–112

Kostopoulos G, Veronikis DK, Efthimiou I (1987) Caffeine blocks absence seizures in the tottering mutant mouse. Epilepsia 28(4): 415–420

Kostopoulos G, Psarropoulou C, Haas H (1988) Membrane properties, response to amines and to tetanic stimulation of hippocampal neurons in the genetically epileptic mutant mouse tottering. Exp Brain Res 72: 45–50

Krnjevic K (1983) GABA — mediated inhibitory mechanisms in relation to epileptic discharges. In: Jasper HH, Van Gelder NM (eds) Basic mechanisms of neuronal hyperexcitability. Alan R Liss, New York, pp 249–280

Levitt P, Noebels JL (1981) Mutant mouse tottering: selective increase of locus coeruleus axons in a defined single-locus mutation. Proc Natl Acad Sci U.S.A. 78: 4630–4634

Levitt P, Law C, Pylypiw A, Ross LL (1984) Central adrenergic receptors in the inherited noradrenergic hyperinnervated mutant mouse tottering. Neurosci Abstr 10: 179

Lopes da Silva FH (1987) Hippocampal kindling: physiological evidence for progressive inhibition. Adv Epileptol 16: 57–62

Madison DV, Nicoll RA (1988) Noradrenaline decreases synaptic inhibition in the rat hippocampus. Brain Res 442: 131–138

Malouf AT, Robbins CA, Schwartzkroin PA (1990) Epileptiform activity in hippocampal slice cultures with normal inhibitory synaptic drive. Neurosci Lett 108: 76–80

McCarren M, Alger BE (1985) Use-dependent depression of IPSPs in rat hippocampal pyramidal cells in vitro. J Neurophysiol 53: 557–571

McIntyre DC, Wong RKS (1986) Cellular and synaptic properties of amygdala-kindled pyriform cortex in vitro. J Neurophysiol 55: 1295–1307

Meldrum BS, Croucher MJ, Badman C, Collins JS (1983) Antiepileptic action of excitatory amino acid antagonists in the photosensitive baboon, Papio papio. Neurosci Lett 39: 101–104

Mody I, Stanton PK, Heinemann U (1988) Activation of N-methyl-D-aspartate receptors parallels changes in cellular and synaptic properties of dentate gyrus granule cells after kindling. J Neurophysiol 59: 1033–1054

Newberry NR, Nicoll RA (1984) A bicucculine-resistant inhibitory post-synaptic potential in rat hippocampal pyramidal cells in vitro. J Physiol (Lond) 348: 239–254

Noebels JL, Sidman RL (1979) Inherited epilepsy: spike-wave and focal motor seizures in the mutant mouse tottering. Science 204: 1334–1336

Noebels JL (1984) A single gene error of noradrenergic axon growth synchronizes central neurons. Nature 310: 409–411

Noebels JL, Rutecki PA (1990) Altered hippocampal network excitability in the hypernoradrenergic mutant mouse tottering. Brain Res 524: 225–230

Olpe H-R (1982) The locus coeruleus as a target for the activating action of vincamine, nicotine and caffeine. Experientia 38: 757

Poolos NP, Mauk MD, Kocsis JD (1987) Activity-evoked increases in extracellular potassium modulate presynaptic excitability in the CA_1 region of the hippocampus. J Neurophysiol 58: 404–416

Prince DA, Connors BW (1986) Mechanisms of interictal epileptogenesis. In: Delgado-Esqueta A, et al (eds) Advances in neurology, vol 44. Raven, New York, pp 275–299

Psarropoulou C, Kostopoulos G (1990) Long term enhancement of post synaptic excitability after brief exposure to Mg^{2+} free medium in normal and epileptic mice. Brain Res 508: 70–75

Psarropoulou C, Angelatou F, Matsokis N, Veronikis DK, Kostopoulos G (1987) Absence of modification in GABA and benzodiazepine binding and in choline acetyltransferase activity in brain areas of the epileptic mutant mouse tottering. Gen Pharmacol 18(6): 593–597

Schwartzkroin PA, Prince DA (1980) Changes in excitatory and inhibitory synaptic potentials leading to epileptogenic activity. Brain Res 183: 61–76

Shefner SA, Chiu TH (1986) Adenosine inhibits locus coeruleus neurons: an intracellular study in a rat brain slice preparation. Brain Res 366: 364–368

Somjen GG (1979) Extracellular potassium in the central nervous system. Ann Rev Physiol 41: 159–177

Spencer WA, Kandel ER (1969) Synaptic inhibition in seizures. In: Jasper HH, Ward AA, Pope A (eds) Basic mechanisms of the epilepsies. Little, Brown and Co, Boston, pp 575–603

Stanfield BB (1989) Excessive intra- and supragranular mossy fibers in the dentate gyrus of tottering (tg/tg) mice. Brain Res 480: 294–299

Tancredi V, Avoli M (1987) Control of spontaneous epileptiform discharges by extracellular potassium. An "in vitro" study in the CA_1 subfield of the hippocampal slice. Exp Brain Res 67: 363–372

Taylor-Courval D, Gloor P (1984) Behavioural alterations associated with generalized spike and wave discharges in the EEG of the cat. Exp Neurol 83: 167–186

Traynellis SF, Dingledine R (1988) Potassium-induced spontaneous electrographic seizures in the rat hippocampal slice. J Neurophysiol 59(1): 259–276

Tuff LP, Racine RJ, Adamec R (1983) The effect of kindling on GABA-mediated inhibition in the dentate gyrus of the rat. I. Paired -pulse depression. Brain Res 277: 79–90

Vornov JJ, Sutin J (1986) Noradrenergic hyperinnervation of motor trigeminal nucleus: alterations in membrane properties and response to synaptic input. J Neurosci 6(1): 30–37

    Authors' address: Dr. G. Kostopoulos, Department of Physiology, University of Patras Medical School, Patras, Greece 26110

J Neural Transm (1992) [Suppl] 35: 125–139

The inhibitory control of the substantia nigra over generalized non-convulsive seizures in the rat

A. Depaulis

Laboratoire de Neurophysiologie et Biologie des Comportements,
Centre de Neurochimie du CNRS, Strasbourg, France

Summary. A system exerting inhibitory control over generalized epilepsies and involving neurons from the substantia nigra has been described by several authors in experimental models of convulsive seizures. In the present study, the existence of such a control system governing absence epilepsy was investigated using models of non-convulsive seizures in the rat. Activation of the GABAergic neurotransmission within the substantia nigra by local injection of GABA agonists (muscimol, THIP) or an inhibitor of GABA degradation (gamma-vinyl GABA) suppresses generalized non convulsive seizures, whether they are genetically determined or induced by systemic injections of gamma-butyrolactone (100 and 200 mg/kg), pentylenetetrazole (20 mg/kg) or THIP (7.5 mg/kg). The ascending dopaminergic nigral output or the GABAergic fibres to the ventromedial thalamus are not critically involved in this control system. By contrast, the GABAergic nigro-collicular pathway appears crucial: bilateral lesion of the superior colliculus abolishes the anti-epileptic effects of intranigral injection of muscimol and blockade of the GABAergic transmission within the superior colliculus results in a suppression of generalized non-convulsive seizures. Finally, activation of collicular cell bodies by low doses of kainic acid significantly suppresses absence seizures. These results suggest the existence of a control system inhibiting generalized non-convulsive seizures which is activated by the release of the tonic inhibition exerted by the nigral GABAergic fibres on collicular neurons. The similarities between this system and the control system described for convulsive seizures are discussed.

Introduction

During the last 10 years, several studies have suggested that the neurotransmission involving gamma-aminobutyric acid (GABA), as well as other inhibitory neurotransmissions, is involved, at the level of the substantia nigra (SN), in the control of generalized convulsive seizures (for review see

Gale, 1985). This structure receives important GABAergic inputs, mainly from the forebrain (e.g. Smith and Bolam, 1989) and shows a high density of GABA receptors as well as one of the highest levels of both endogenous and extracellular GABA content (see Mugnaini and Oertel, 1985). Endogenous GABA levels were found to be decreased in the SN preliminary to seizures induced either by methoxypyridoxine or amygdala kindling (Nitsch and Okada, 1976; Löscher and Schwark, 1985). In 1980, Gale and Iadarola (1980) demonstrated that the anticonvulsant effects of systemic injections of gamma-vinyl GABA and valproate were correlated with an increase in endogenous GABA, more specifically at the nerve terminal level in the SN. In the mongolian gerbil susceptible to convulsive seizures, a lower density of GABA-benzodiazepine receptors was observed in the SN as compared to non susceptible animals (Olsen et al., 1985). Finally, pharmacological potentiation of GABAergic transmission within the SN, by bilateral microinjections of either muscimol, a GABA agonist, or gamma-vinyl GABA, an inhibitor of GABA transaminase, suppresses seizures in different models of generalized convulsive seizures in the rat (Garant and Gale, 1986; Gonzalez and Hettinger, 1984; Iadarola and Gale, 1982; Le Gal La Salle et al., 1983; Löscher et al., 1987; McNamara et al., 1983; Miller et al., 1987; Mirski et al., 1986; Okada et al., 1989; Sperber et al., 1989; Toussi et al., 1987; Turski et al., 1986; Zhang et al., 1991).

Taken together, these data suggest that a GABAergic system in the SN is involved in a control mechanism which, when activated, exerts an inhibitory influence on generalized convulsive seizures (Gale, 1985). When the present study was initiated in 1985, no data were available on the possible involvement of such a nigral inhibitory control over generalized *non-convulsive* seizures (GNCS). As indicated in the present volume (see Marescaux; Vergnes), this type of seizure clearly involves different neural mechanisms from those associated with convulsive seizures. In particular, systemic injection of GABA mimetics results in aggravation of the seizures in both animals and man. The aim of the present study was thus 1) to examine the possibility that generalized non-convulsive seizures may also be sensitive to a nigral control, as described in models of generalized convulsive seizures, and 2) to characterize the output pathways involved in this control mechanism.

Materials and methods

Male Wistar rats (350–400 g) were used in all experiments in this study and were chosen from either the Genetic Absence Epilepsy Rats from Strasbourg (GAERS) strain selected in our laboratory, or the control strain (non-epileptic rats; see Marescaux, this volume). All rats were kept in individual cages under a 12/12 h normal light/ dark cycle, with food and water ad libitum.

All animals were implanted under general anaesthesia (ketamine hydrochloride, 100 mg/kg i.p.) as previously described (Depaulis et al., 1988, 1989, 1990d), with four stainless-steel electrodes placed bilaterally over the frontal and parietal cortex and connected to a micro-connector and bilateral stainless-steel cannulae (o.d., 0.4 mm; i.d., 0.3 mm) aimed at the structure under investigation. Electrolytic lesions were produced stereotaxically by using a steel electrode (o.d., 0.125 mm) insulated with epoxy to within 0.5–1.0 mm of the tip. The lesions were induced by passing cathodal current of 2 mA for 20 s at several sites for a given structure (see Depaulis et al., 1990d). Selective lesions of dopaminergic neurons of the SN were effected by bilateral injections of 6-hydroxydopamine (5 µg/side in 1 µl), using chronically implanted guide cannulae aimed at the SN and the intracerebral injection technique (see below). Hemisection of the medial forebrain bundle was performed using a retractable knife technique, as described by Sclafani and Grossman (1969). The knife consisted of a stainless-steel guide cannula (o.d.: 0.3 mm; i.d.: 0.15 mm) bent at one end in such a way that a 0.13 mm stainless-steel wire forced through the cannula would extend in the direction of the curved end (see Depaulis et al., 1990d).

Intracerebral injections were made in the awake animal, as described previously (Depaulis et al., 1988, 1989, 1990a), by using stainless-steel cannulae (o.d.: 0.28 mm, i.d.: 0.18 mm). Bilateral injections of the drug or the vehicle were given simultaneously with a volume of 200 nl per side. Rats were injected a maximum of 5 times in a randomized order for drug conditions. The electroencephalograms (EEG) were recorded in the freely moving animal from ipsilateral fronto-parietal electrodes. On the day of experiment, after 15 min of habituation to the test cage, the EEG was recorded for a 20 min reference period. When a drug was injected, the EEG was then recorded continuously for up to two hours.

Upon completion of an experiment, the animals were killed by an overdose of pentobarbital (100 mg/kg, i.p.). Brains were removed from the skull, hardened in formaldehyde (3%) and cut in 20 µm coronal sections stained with Cresyl violet. Each injection site was localized with reference to the atlas of Paxinos and Watson (1982). When biochemical assays were performed, the brain was sectioned vertically at the level of the optic chiasma. The caudal part was kept for histological control, whereas the rostral part was frozen in liquid nitrogen for assays of noradrenaline, dopamine and serotonin. Noradrenaline, dopamine and serotonin were determined simultaneously using a reversed-phase high-pressure liquid chromatography (HPLC) procedure coupled with electrochemical detection (see Depaulis et al., 1990d).

Data on generalized non-convulsive seizures were expressed as mean ± S.E.M. cumulative duration of spike-and-wave discharges (s) for 20 min periods. Within each period, the means between test conditions were compared using a non-parametric analysis of variance (Siegel, 1956). Paired comparisons versus control conditions (vehicle only) or between treatments were made using non-parametric tests for related (Wilcoxon) or independent (Mann-Whitney) samples, according to the experiment. Data on biochemical assays were compared between groups using the Mann-Whitney test.

Results

Involvement of the substantia nigra in the control of generalized non-convulsive seizures

The involvement of the SN in the control of GNCS was suggested by the complete suppression of spike-and-wave discharges observed in rats with spontaneous absence seizures following bilateral injections of muscimol at doses as low as 1 and 2 ng/side (Depaulis et al., 1988). This suppression was significant for 60–80 min as compared to the results after bilateral injection

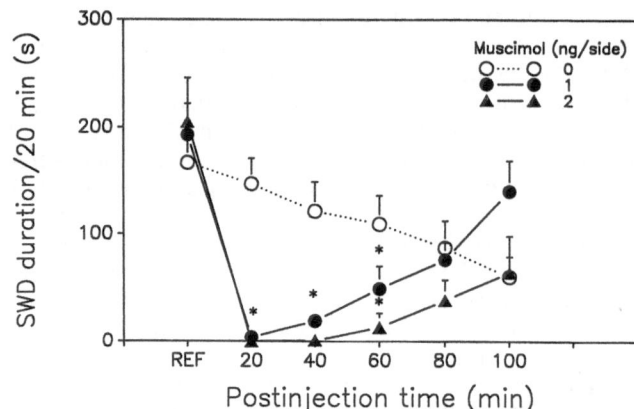

Fig. 1. Mean ± S.E.M. cumulative duration of spike-and-wave discharges (SWD) per 20 min period before (REF) and after a bilateral injection of muscimol into the substantia nigra. *p < 0.02, Wilcoxon test, N = 7

of saline at the same sites (Fig. 1). At these doses, no behavioural modifications (e.g. increased activity, stereotypies) were observed. This effect was observed only after *bilateral* injection of muscimol; a unilateral injection had no effect on the seizures. Bilateral injection of GABA antagonists (bicuculline methiodide, SR 42641) into the SN also failed to modify the SWD at doses without behavioural effects (Depaulis et al., 1988).

Following bilateral injection of muscimol into the SN, suppression of SWD was observed either upon cortical recording or upon EEG recording within the ventro-lateral thalamus using depth electrodes (Depaulis et al., 1988). These structures have been suggested to be the critical site of generation of GNCS (see Vergnes, this volume). When EEG recording was carried out using bipolar depth electrodes located within the SN, no local paroxysmal activity was observed following intranigral muscimol (Depaulis et al., 1989).

The suppressive effect of intranigral muscimol on SWD was confirmed in three pharmacological models of GNCS in the rat. In rats from the non-epileptic strain, it was possible to induce SWD following systemic injection of gamma-butyrolactone (100 and 200 mg/kg), low doses of pentylenetetrazole (PTZ; 20 mg/kg) or THIP (7.5 mg/kg) as described previously (see Marescaux; Snead, this volume). In all three models, intranigral bilateral injections of muscimol (2 ng/side) significantly reduced the SWD (Depaulis et al., 1989).

Involvement of the GABAergic transmission

The initial results suggested that bilateral activation of GABAergic receptors within the SN resulted in suppression of GNCS. The GABAergic nature of

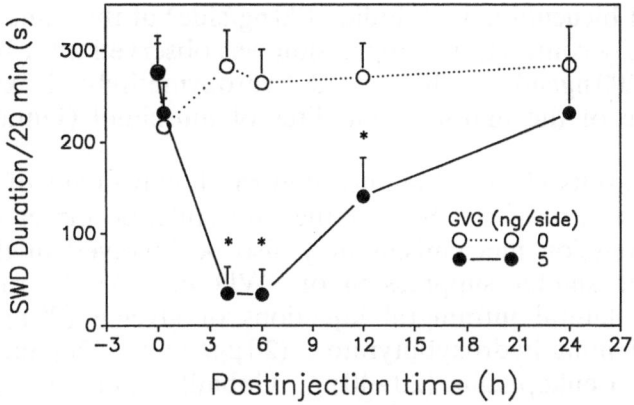

Fig. 2. Mean ± S.E.M. cumulative duration of spike-and-wave discharges (SWD) per 20 min period before (REF) and after a bilateral injection of gamma-vinyl GABA (GVG) into the substantia nigra. *$p < 0.05$, Wilcoxon test, N = 3

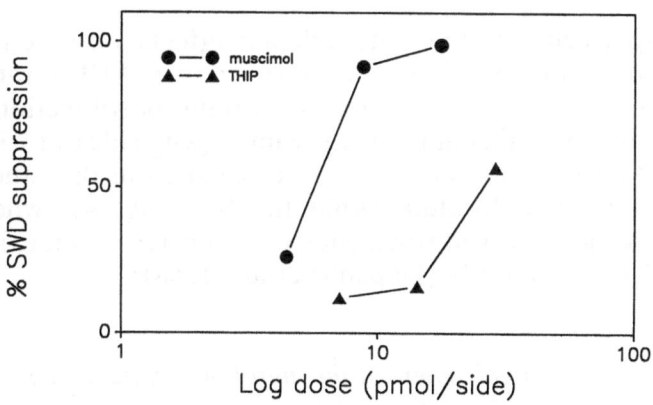

Fig. 3. Dose-response effects of bilateral intranigral injections of muscimol and THIP on the percentage of suppression of spike-and-wave discharges during 40 min postinjection, as compared to control conditions (saline). N = 7 and 8 for muscimol and THIP, respectively

the mechanisms described above was further confirmed by the following data. The suppression of SWD in GAERS rats observed after intranigral muscimol was reproduced by bilateral injection of gamma-vinyl GABA (GVG; 5 ug/side). In this case, the suppression of the seizures was significant 4 to 12 h after the injection, a time-course which is consistent with the increase in GABA content following GVG administration (Fig. 2; Depaulis, 1990c).

The suppressive effects of intranigral muscimol were likewise found to be (1) dose-dependent (Fig. 3); (2) reproduced by THIP, another GABA$_A$-agonist, although less effectively (Fig. 3), as well as by l-baclofen (20 ng/side), a GABA$_B$ agonist (Depaulis, 1990c) and (3) reversed by a subsequent injection of bicuculline methiodide (20 ng/side) at the same site (Depaulis et al., 1988). By contrast, no suppression was observed after bilateral injection of up to 200 ng/side of hydroxy-3 hydroxymethyl-5 isoxazole (Depaulis, 1990c), one of the major metabolites of muscimol (Unnerstall and Pizzi, 1981).

These results clearly demonstrated the involvement of the GABAergic transmission, within the SN, in the anti-epileptic effect observed. Other neurotransmission mechanisms may also be involved in this anti-epileptic effect, since similar suppression of SWD in GAERS rats was obtained following bilateral intranigral injections of glycine (20 µg/side; Depaulis, 1990c), gamma-hydroxybutyrate (20 µg/side; Depaulis, 1990c) or (D-Ala^2)Met-enkephalin (20–40 µg; Gobaille, personal communication). However, for these compounds, high doses were necessary to obtain significant suppression of GNCS.

Site-specificity of the substantia nigra

The site-specificity of the anti-epileptic effects observed previously was indicated by a mapping study performed in GAERS rats, in which the suppressive effects on SWD during 40 min postinjection were assessed following bilateral injections of muscimol (2 ng/side) at different levels of the ventral part of the midbrain. It was clear from this study that the most effective sites were located within the SN (Fig. 4), whereas almost no suppression was observed when sites were located further than 0.5 mm from the boundaries of the SN (Depaulis et al., 1988).

Involvement of the nigral output pathways

The data collected during the first part of this study demonstrated that activation of the GABAergic transmission within the SN results in a suppression of SWD. Several data from our laboratory indicated that the SN is not directly involved in the genesis or even the propagation of SWD. In this structure, only irregular SWD of low amplitude were recorded in curarized GAERS rats (Vergnes et al., 1990b) or in chronically implanted non-epileptic rats given either 20 mg/kg of PTZ or 100 mg/kg of GBL (Depaulis, 1990c). Furthermore, a bilateral electrolytic lesion of the SN did not abolish the SWD (Depaulis, 1990c) in contrast with the complete suppression of SWD observed following exclusion of the fronto-parietal cortex or lesions of the lateral thalamus (Vergnes et al., 1990a). These results suggested that the anti-epileptic effects on GNCS observed after bilateral injection of muscimol into the SN were not secondary to a functional exclusion of the structure. On the contrary, as was suggested in the case of generalized convulsive seizures (Gale, 1985), the activation of nigral GABAergic transmission very likely involved other structures via one

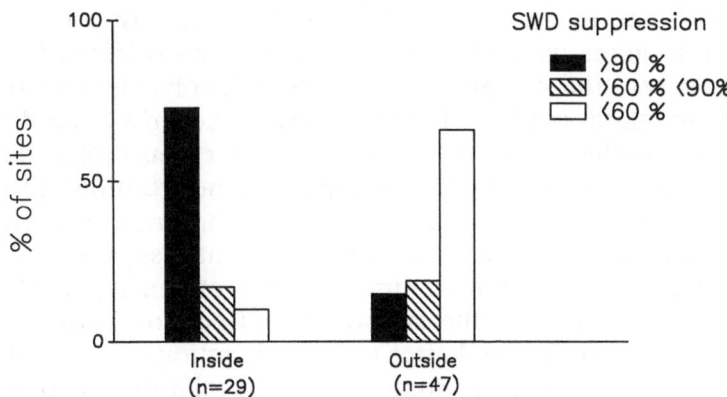

Fig. 4. Percentage of injection sites inside and outside the limits of the substantia nigra according to the percentage of spike-and-wave suppression, as compared to control conditions (saline), during 40 min after a bilateral injection of muscimol (2 ng/side)

or several nigral output pathways. An important question was then to determine the critical nigral efferent pathways involved.

The SN has several efferent pathways which originate from two main areas. First, dopaminergic cell bodies are located in the pars compacta and project primarily to the neostriatum and foreparts of the brain through the medial forebrain bundle (e.g., Andén et al., 1984; Bentivoglio et al., 1979). Activation of GABAergic receptors in the SN pars reticulata appears to result in stimulation of dopaminergic cells through a disinhibition process (e.g., Grace and Bunney, 1979). Secondly, histological, electrophysiological and pharmacological studies have demonstrated the existence of three main GABAergic nigral projections with cell bodies localized in the pars reticulata of the SN and with nerve terminals ending within (1) the inter-mediate and deep layers of the superior colliculus (e.g., Bentivoglio et al., 1979); (2) the ventromedial thalamic nucleus (e.g., Di Chiara et al., 1979; Kilpatrick et al., 1980; McLeod et al., 1980), and (3) the pedunculo-pontine nucleus of the tegmentum (i.e., nigro-tegmental projection; Childs and Gale, 1983; Hopkins and Niessen, 1976). The thalamic and collicular pro-jections appear to exert a tonic inhibition on the respective target cells (Chevalier et al., 1981, 1985; McLeod et al., 1980). Furthermore, activation of GABAergic receptors within the SN results in inhibition of these two GABAergic outputs (Chevalier et al., 1985; McLeod et al., 1980).

Dopaminergic output

Data from our laboratory strongly suggested the critical involvement of the dopaminergic output in the nigral control of GNCS. Systemic adminis-tration of dopaminergic agonists significantly suppressed SWD in GAERS animals, whereas injections of antagonists increased the occurrence of the

seizures (Warter et al., 1988). Furthermore, neurotoxic lesions of the
dopaminergic fibres were shown to aggravate the seizures (Lannes et al.,
· 1989). However, the evidence for a critical involvement of dopaminergic
output in the nigral control of GNCS was weakened by the data from two
experiments conducted in GAERS rats. After neurotoxic destruction of
nigral dopaminergic cells by 6-hydroxydopamine, resulting in depletion of
more than 60% of the dopamine levels in the forebrain, bilaterally injected
muscimol (2 ng/side) into the SN still significantly suppressed SWD, compared to either sham-operated animals or the reference period (Depaulis et
al., 1990d). Furthermore, the suppressive effects of intranigral muscimol
(2 ng/side) were still obtained after unilateral section of the medial forebrain
bundle, which resulted, in some cases, in almost complete unilateral
dopamine depletion in the forebrain (Depaulis et al., 1990d).

The nigrothalamic pathway

The involvement of the GABAergic output pathway was investigated 1) by
testing whether intranigral muscimol was still suppressive on the SWD after
bilateral electrolytic lesions of each projection area, and 2) by examining
the effects of blocking GABAergic transmission within these regions. Using
these two approaches, it appeared rather clearly that the nigrothalamic
pathway is not involved in the nigral control of the SWD. It was still
possible to suppress SWD in GAERS rats significantly by intranigral
muscimol (2 ng/side) after extensive bilateral electrolytic lesionning of the
ventromedial nucleus of the thalamus (Depaulis et al., 1990d). Furthermore, blockade of GABAergic transmission within the ventromedial
nucleus of the thalamus using bilateral micro-injections of a GABA
antagonist, picrotoxin (10 ng/side), failed to suppress SWD significantly
(Depaulis et al., 1990d).

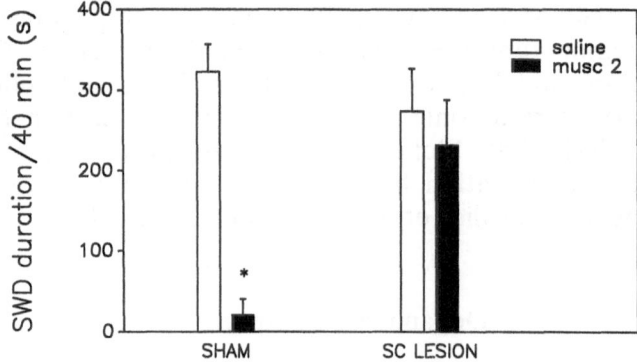

Fig. 5. Mean ± S.E.M. cumulative duration of spike-and-wave discharges during
40 min after a bilateral injection of saline or muscimol (2 ng/side; musc 2) into the
substantia nigra in control (SHAM; N = 7) and SC-lesioned rats (SC LESION; N = 6).
*p < 0.01 vs saline, Wilcoxon test

The nigrocollicular pathway

Contrary to what has been observed for the thalamic outputs, several data suggested the critical involvement of the nigrocollicular GABAergic pathway in the nigral control over GNCS described above. Bilateral lesions of the superior colliculus including the superficial, intermediate and deep layers, antagonized the anti-epileptic effects resulting from bilateral injections of muscimol (2 ng/side) into the SN, as compared to sham-operated animals (Fig. 5; Depaulis et al., 1990d). Moreover, blockade of GABAergic transmission within the deep layers of the superior colliculus, by bilateral injection of picrotoxin (20 and 40 ng/side), resulted in a significant suppression of SWD for 40 min, compared to saline injection (Depaulis et al., 1990a,d). No behavioural modifications were noted in the animals injected with picrotoxin in the superior colliculus, and this suppression was reproduced by bilateral injection of bicuculline methiodide (5 ng/side) into the same region. It is interesting that *unilateral* injection of 40 ng of picrotoxin did not suppress SWD. Muscimol injected bilaterally (up to 80 ng/side) into the superior colliculus likewise failed to modify the occurrence of SWD (Depaulis et al., 1990a).

These last-mentioned results, together with data from the literature (Chevalier et al., 1985), suggested that the anti-epileptic effects induced by intranigral activation of GABAergic transmission were due to *activation* of collicular neurons secondary to inhibition of GABAergic nigrocollicular fibres. This is further supported by the fact that excitation of cell bodies by bilateral injections of low doses of kainate (4 and 8 ng/side) into the superior colliculus significantly suppressed SWD in GAERS (Fig. 6; Depaulis et al., 1990d). Again, during this suppression, the animals showed no signs of hyperactivity or agitation, and their basal EEG was very similar to control tracings. These results are in agreement with a previous report

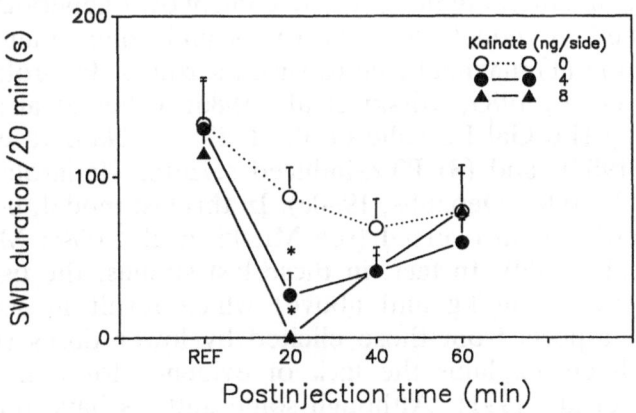

Fig. 6. Mean ± S.E.M. cumulative duration of spike-and-wave discharges per 20 min periods before (REF) and after a bilateral injection of kainate into the superior colliculus. *$p < 0.01$ vs saline, Wilcoxon test, N = 7

according to which bilateral intracollicular injections of glutamate suppressed SWD induced by systemic injection of low doses of PTZ (Redgrave et al., 1988).

Discussion

The results of this study clearly indicate the existence, within the central nervous system, of a control mechanism exerting an inhibitory influence on GNCS. More precisely, they suggest that this inhibitory control system involves GABAergic neurons with cell bodies localized in the pars reticulata of the SN and with nerve terminals ending in the superior colliculus. According to data from the literature (Chevalier et al., 1985), collicular neurons receive a potent inhibitory input from the SN. This inhibition can be suppressed by activation of GABAergic transmission within the SN or by its blockade at the level of the superior colliculus. The disinhibition thus produced results in an increase of the firing of collicular neurons which, in turn, suppresses GNCS. As suggested by different data obtained in this study, this disinhibition process results in anti-epileptic effects only when it occurs bilaterally: unilateral pharmacological manipulations have no effect.

The SN or the superior colliculus do not appear to be critical in the genesis of the SWD in absence epilepsy. Lesioning of these structures does not interfere with the occurrence of seizures — in contrast to lesions of such structures as the lateral part of the thalamus or the fronto-parietal cortex (see Vergnes, this volume) — and only weak SWD are recorded there from depth electrodes. The suppressive effects on GNCS observed during this study thus cannot be explained by a "functional exclusion" of the SN or the superior colliculus. The inhibitory control system described above thus rather appears to be a "gate-control" mechanism, the actual suppression of SWD taking place in brain structures distant from the SN.

The existence of a nigral inhibitory gate control over generalized seizures was first described in models of convulsions and, since then, has been mainly confirmed in (1) maximal electro-shock seizures (Iadarola and Gale, 1982; De Sarro et al., 1985; Mirski et al., 1986; Miller et al., 1987); (2) amygdala kindling (Le Gal La Salle et al., 1983; Mc Namara et al., 1983; Löscher et al., 1987), and (3) PTZ-induced seizures (Iadarola and Gale, 1982; Okada et al., 1989; Depaulis, 1990c). In this last model, however, the results are somewhat controversial (see Mirski et al., 1986; Miller et al., 1987; Zhang et al., 1989). In fact, in these last studies, the use of higher doses of PTZ (i.e. 70 mg/kg and above) which result in seizures with different clinical aspects from those elicited by lower doses (i.e. around 40 mg/kg) very likely explains the lack of evidence for a nigral control (see also Zhang et al., 1991). Although some authors have reported data suggesting the involvement of nigral inhibitory control on audiogenic seizures (Gonzalez and Hettinger, 1984), other data suggest that this form of convulsive seizure is not controlled by such a mechanism (Frye et al.,

1983; Depaulis et al., 1990b). In contrast to most models of generalized convulsive epilepsy, audiogenic seizures in the rat involve mainly brainstem structures (see Depaulis et al., 1990b).

Although the neural mechanisms involved in absence epilepsy are, at least in part, very different from those causing generalized convulsive epilepsies — especially in regard to GABAergic transmission — there is a great similarity between the nigral control system described in the present study and the system described in the various reports on studies using models of convulsive seizures. Furthermore, in both cases, a bilateral pharmacological manipulation is necessary to affect the occurrence of the seizures (Iadarola and Gale, 1982; Le Gal La Salle et al., 1983; Turski et al., 1986) and, apart from flurothyl-induced seizures (Sperber et al., 1989), blockade of GABAergic transmission in the SN had no effects on either convulsive seizures (Iadarola and Gale, 1982; Le Gal La Salle et al., 1983) or GNCS. Very few studies have so far investigated the involvement of the different nigral outputs in the control of convulsive seizures. In the models in which it has been examined (Garant and Gale, 1987; Platt et al., 1987), the hypothesis of an involvement of the dopaminergic nigral output pathway has been rejected. Some data have also indicated that the nigrothalamic GABAergic pathway is not involved in the nigral control of maximal electro-shock seizures (Garant and Gale, 1987), and lesioning of the ventromedial thalamus region does not modify susceptibility to flurothyl-induced tonic-clonic seizures in the rat (Moshé et al., 1985). The lack of involvement of both the dopaminergic and the GABAergic nigrothalamic projections in the nigral inhibitory control over GNCS, as reported in the present study, is thus in keeping with conclusions reached in these previous studies performed in models of convulsive seizures.

By contrast, the present data lead to the same conclusions as a similar study using maximal electro-shock seizures, in which the anticonvulsant effects of intranigral muscimol were selectively abolished by lesioning of the superior colliculus (Garant and Gale, 1987). The integrity of the nigrocollicular GABAergic pathway thus appears critical in the inhibitory effects observed following activation of GABAergic transmission in the SN in both forms of generalized seizures. This hypothesis is further confirmed by the fact that blockade of GABAergic transmission in the superior colliculus suppresses convulsive seizures (Dean and Gale, 1989) as well as GNCS. As yet no further data have been reported concerning the possible anticonvulsant effects of selective activation of cell bodies in the superior colliculus in animal models of generalized convulsive seizures. It is possible that different populations of neurons within the superior colliculus are involved in the inhibitory control of convulsive and non-convulsive seizures.

In conclusion, the data reported in the present study demonstrate the existence, within the central nervous system, of a inhibitory control mechanism involving GABAergic transmission of the SN. They strongly suggest that this system acts as a gate-control mechanism in which the GABAergic nigrocollicular pathway plays a critical role. Inhibition of the tonic activity of this pathway, either at the cell body or at the nerve ending level, results in an increased activity of some collicular neurons. By way of some yet unknown mechanism, this disinhibition of collicular neurons suppresses SWD. Since forebrain structures are involved in the generation of SWD (see Vergnes, present volume), it is very likely that ascending, rather than descending, projections from the superior colliculus are involved in this control. The results of the present study are very similar to data obtained in some models of generalized convulsive seizures and suggest that a similar nigral inhibitory control of generalized seizures exists, at least for some forms of generalized non-convulsive or convulsive epilepsies.

Acknowledgements

We are thankful to A. Boehrer, Z. Liu, and C. Schleef for their valuable technical assistance. This work was supported by grants from I.N.S.E.R.M. (Contrat de Recherche Externe n° 866017) and from the "Fondation pour la Recherche médicale".

References

Andén NE, Carlsson A, Dahlström A, Fuxe K, Hillarp NA, Larsson K (1984) Demonstration and mapping out of nigro-neostriatal dopamine neurons. Life Sci 3: 523–530

Bentivoglio M, Van der Kooy D, Kuypers HGJM (1979) The organization of the efferent projections of the substantia nigra in the rat. A retrograde fluorescent double labeling study. Brain Res 174: 1–17

Chevalier G, Thierry AM, Shibazaki T, Feger J (1981) Evidence for a GABAergic inhibitory nigro-collicular pathway in the rat. Neurosci Lett 21: 67–70

Chevalier G, Vacher S, Deniau JM, Desban M (1985) Disinhibition as a basic process in the expression of striatal functions. The striato-nigral influence on tecto-spinal/tecto-diencephalic neurons. Brain Res 334: 215–226

Childs JA, Gale K (1983) Neurochemical evidence for a nigrotegmental GABAergic projection. Brain Res 258: 109–114

Dean P, Gale K (1989) Anticonvulsant action of GABA receptor blockade in the nigro-collicular target region. Brain Res 477: 391–395

Depaulis A, Vergnes M, Marescaux C, Lannes B, Warter JM (1988) Evidence that activation of GABA receptors in the substantia nigra suppresses spontaneous spike-and-wave discharges in the rat. Brain Res 448: 20–29

Depaulis A, Snead OC, III, Marescaux C, Vergnes M (1989) Suppressive effects of intranigral injection of muscimol in three models of generalized non-convulsive epilepsy induced by chemical agents. Brain Res 498: 64–72

Depaulis A, Liu Z, Vergnes M, Marescaux C, Micheletti G, Warter JM (1990a) Suppression of spontaneous generalized non-convulsive seizures in the rat by microinjection of GABA antagonists into the superior colliculus. Epilepsy Res 5: 192–198

Depaulis A, Marescaux C, Liu Z, Vergnes M (1990b) The GABAergic nigro-collicular pathway is not involved in the inhibitory control of audiogenic seizures in the rat. Neurosci Lett 111: 269–274

Depaulis A (1990c) Etude neuropharmacologique du contrôle inhibiteur par la substance noire des crises d'épilepsie généralisée non-convulsive chez le rat. Ph.D. dissertation, Louis Pasteur University, Strasbourg

Depaulis A, Vergnes M, Liu Z, Kempf E, Marescaux C (1990d) Involvement of the nigral output pathways in the inhibitory control of the substantia nigra over generalized non-convulsive seizures in the rat. Neuroscience 39: 339–349

De Sarro G, Meldrum BS, Reavill C (1985) Anticonvulsant action of 2-amino-7-phosphonoheptanoic acid in the substantia nigra. Eur J Pharmacol 106: 175–179

Di Chiara G, Porceddu ML, Morelli M, Mulas ML, Gessa GL (1979) Evidence for a GABAergic projection from the substantia nigra to the ventromedial thalamus and to the superior colliculus of the rat. Brain Res 176: 273–284

Frye GD, McCown TJ, Breese GR (1983) Characterization of susceptibility to audiogenic seizures in ethanol-dependent rats after microinjection of gamma-aminobutyric acid (GABA) agonists into the inferior colliculus, substantia nigra and medial septum. J Pharmacol Exp Ther 227: 663–670

Gale K (1985) Mechanisms of seizure control mediated by gamma aminobutyric acid: role of the substantia nigra. Fed Proc 44: 2414–2424

Gale K, Iadarola MJ (1980) Seizure protection and increased nerve-terminal GABA: delayed effects of GABA transaminase inhibition. Science 208: 288–291

Garant DS, Gale K (1986) Intranigral muscimol attenuates electrographic signs of seizure activity induced by intravenous bicuculline in rats. Eur J Pharmacol 124: 365–369

Garant DS, Gale K (1987) Substantia nigra-mediated anticonvulsant actions: role of nigral output pathways. Exp Neurol 97: 143–159

Gonzalez LP, Hettinger MK (1984) Intranigral muscimol suppresses ethanol withdrawal seizures. Brain Res 298: 163–166

Grace AA, Bunney BS (1979) Paradoxical GABA excitation of nigral dopaminergic cells: indirect mediation through reticulata inhibitory neurons. Eur J Pharmacol 59: 211–218

Hopkins DA, Niessen LW (1976) Substantia nigra projections to the reticular formation, superior colliculus and central gray in the rat, cat and monkey. Neurosci Lett 2: 253–259

Iadarola MJ, Gale K (1982) Substantia nigra: site of anti-convulsant activity mediated by gamma-aminobutyric acid. Science 218: 1237–1240

Kilpatrick IC, Starr MS, Fletcher A, James TA, MacLeod NK (1980) Evidence for a GABAergic nigrothalamic pathway in the rat. I. Behavioral and biochemical studies. Exp Brain Res 40: 45–54

Lannes B (1989) Réflexion sur les relations entre maladie de Parkinson et Epilepsies. A propos des effets de lésions de neurones catécholaminergiques dans un modèle d'épilepsie généralisée non-convulsive du rat. Medical Doctorate dissertation, Strasbourg

Le Gal La Salle G, Kaijima M, Feldblum S (1983) Abortive amygdaloid kindled seizures following microinjection of gamma-vinyl-GABA in the vicinity of substantia nigra in rats. Neurosci Lett 36: 69–74

Löscher W, Schwark WS (1985) Evidence for impaired GABAergic activity in the substantia nigra of amygdaloid kindled rats. Brain Res 339: 146–150

Löscher W, Czuczwar SJ, Jäkel R, Schwarz M (1987) Effect of microinjections of gamma-vinyl GABA or isoniazid into substantia nigra on the development of amydala kindling in rats. Exp Neurol 95: 622–638

McLeod NK, James TA, Kilpatrick IC, Starr MS (1980) Evidence for a GABAergic nigrothalamic pathway in the rat. II. Electrophysiological studies. Exp Brain Res 40: 55–61

McNamara JO, Rigsbee JC, Galloway MT (1983) Evidence that substantia nigra is crucial to neural network of kindled seizures. Eur J Pharmacol 86: 485–486

Miller JW, McKeon AC, Ferrendelli JA (1987) Functional anatomy of pentylenetetrazol and electroshock seizures in the rat brainstem. Ann Neurol 22: 615–621

Mirski MA, McKeon AC, Ferendelli JA (1986) Anterior thalamus and substantia nigra: two distinct structures mediating experimental generalized seizures. Brain Res 397: 377–380

Moshé SL, Okada R, Albala BJ (1985) Ventromedial thalamic lesions and seizure susceptibility. Brain Res 337: 368–372

Mugnaini E, Oertel WH (1985) An atlas of the distribution of GABAergic neurons and terminals in the rat CNS as revealed by GAD immunohistochemistry. In: Björklund A, Hökfelt T (eds) Handbook of chemical neuroanatomy, vol IV. GABA and neuropeptides in the CNS. Elsevier, Amsterdam, pp 436–608

Nitsch C, Okada Y (1976) Differential decrease of GABA in the substantia nigra and other discrete regions of the rabbit brain during the preictal period of methoxypyridoxine-induced seizures. Brain Res 105: 173–178

Okada R, Negishi N, Nagaya H (1989) The role of the nigrotegmental GABAergic pathway in the propagation of pentylenetetrazol-induced seizures. Brain Res 480: 383–387

Olsen RW, Wamsley JK, McCabe RT, Lee RJ, Lomax P (1985) Benzodiazepine/gamma-aminobutyric acid receptor deficit in the midbrain of the seizure-susceptible gerbil. Proc Natl Acad Sci USA 82: 6701–6705

Paxinos G, Watson C (1982) The rat brain in stereotaxic coordinates. Academic Press, Sydney

Platt K, Butler LS, Bonhaus DW, McNamara JO (1987) Evidence implicating alpha-2 adrenergic receptors in the anticonvulsant action of intranigral muscimol. J Pharmacol Exp Ther 241: 751–754

Redgrave P, Dean P, Simkins M (1988) Intratectal glutamate suppresses pentylenetetrazole-induced spike-and-wave discharges. Eur J Pharmacol 158: 283–287

Sclafani A, Grossman SP (1969) Hyperphagia produced by knife cuts between the medial and lateral hypothalamus. Physiol Behav 4: 533–538

Siegel S (1956) Nonparametric statistics for the behavioral sciences. McGraw-Hill, New York

Smith Y, Bolam JP (1989) Neurons of the substantia nigra reticulata receive a dense GABA-containing input from the globus pallidus in the rat. Brain Res 493: 160–167

Sperber EF, Wurpel JND, Zhao DY, Moshé SL (1989) Evidence for the involvement of nigral GABA$_A$ receptors in seizures of adult rats. Brain Res 480: 378–382

Toussi HR, Schatz RA, Waszczak BL (1987) Suppression of methionine sulfoximine seizures by intranigral gamma-vinyl GABA injection. Eur J Pharmacol 137: 261–264

Turski L, Cavalheiro EA, Schwarz M, Turski WA, De Moraes Mello EA, Bortolotto ZA, Klockgether T, Sontag K-H (1986) Susceptibility to seizures produced by pilocarpine in rats after microinjection of isoniazid or gamma-vinyl-GABA into the substantia nigra. Brain Res 370: 294–309

Unnerstall JR, Pizzi WJ (1981) Muscimol and gamma-hydroxybutyrate: similar interactions with convulsant agents. Life Sci 29: 337–344

Vergnes M, Marescaux C, Depaulis A, Micheletti G, Warter JM (1990a) Spontaneous spike and wave discharges in Wistar rats: a model of genetic generalized nonconvulsive epilepsy. In: Avoli M, Gloor P, Kostopoulos G, Naquet R (eds) Generalized epilepsy: cellular, molecular and pharmacological approaches. Birkhäuser, Boston, pp 238–253

Vergnes M, Marescaux C, Depaulis A (1990b) Mapping of spontaneous spike and wave discharges in wistar rats with genetic generalized non-convulsive epilepsy. Brain Res 523: 87–91

Warter JM, Vergnes M, Depaulis A, Tranchant C, Rumbach L, Micheletti G, Marescaux C (1988) Effect of drugs affecting dopaminergic neurotransmission in rats with spontaneous petit mal-like seizures. Neuropharmacology 27: 269–274

Zhang H, Rosenberg HC, Tietz EI (1989) Injection of benzodiazepines but not GABA or muscimol into pars reticulata of substantia nigra suppresses pentylenetetrazol seizures. Brain Res 488: 73–79

Zhang H, Rosenberg HC, Tietz EI (1991) Anticonvulsant actions and interaction of GABA agonists and a benzodiazepine in pars reticulata of substantia nigra. Epilepsy Res 8: 11–20

Author's address: Dr. A. Depaulis, L.N.B.C., Centre de Neurochimie du CNRS, 5, rue Blaise Pascal, F-67084 Strasbourg Cedex, France

J Neural Transm (1992) [Suppl] 35: 141–153
© Springer-Verlag 1992

Mapping of cerebral energy metabolism in rats with genetic generalized nonconvulsive epilepsy

A. Nehlig[1], **M. Vergnes**[2], **C. Marescaux**[3], and **S. Boyet**[1]

[1] INSERM U272, Université de Nancy I, Nancy, [2] Centre de Neurochimie, CNRS, Strasbourg, and [3] Groupe de Recherche de Physiopathologie Nerveuse, Clinique Neurologique, Hôpital Civil, Strasbourg, France

Summary. The quantitative 2-[^{14}C]deoxyglucose autoradiographic method was applied to measure local cerebral metabolic rates of glucose (LCMRglc) in a model of genetic petit-mal-like seizures in a strain of Wistar rats. During the experimental period, epileptic rats exhibited synchronous spike-and-wave discharges, whereas the EEG pattern of control animals was normal. Overall, LCMRglc was consistently higher in epileptic rats than in the non-epileptic controls. The increase in LCMRglc was widespread and concerned all cerebral functional systems studied, whether they exhibit spike-and-wave discharges (neocortex and thalamus), or not (limbic system). These results are in good accordance with positron-emission tomography measurements in humans with typical childhood absence epilepsy. There appears to be a lack of anatomical correlation between areas demonstrating hypermetabolism and areas where spike-and-wave discharges are recorded. The administration of 200 mg/kg ethosuximide completely suppressed spike-and-wave discharges in epileptic rats and did not change the EEG pattern in controls. However, LCMRglc were increased to the same extent over control values in epileptic rats whether they were injected with ethosuximide or untreated. By contrast, when epileptic rats were given 2 mg/kg haloperidol, the frequency and the length of spike-and-wave discharges increased, inducing almost a permanent petit-mal status epilepticus. Haloperidol did not change EEG pattern in controls. In haloperidol-treated epileptic rats, LCMRglc decreased to levels comparable to those measured in untreated control rats. In the presence of haloperidol, LCMRglc were similar in both control and epileptic rats. Thus, the diffuse increase in cerebral energy metabolism in epileptic rats as compared to controls is not directly related to the occurrence of spike-and-wave discharges, and may rather be associated with inhibitory mechanisms involved in their termination and suppression, as well as their spread to limbic and motor structures.

142 A. Nehlig et al.

Introduction

Both the 2-[^{14}C]deoxyglucose (2DG) autoradiographic method of Sokoloff
et al. (1977) and the [^{18}F]fluoro-2-deoxyglucose (FDG) method (Phelps et
al., 1979; Reivich et al., 1979) have been extensively used to study changes
in local cerebral metabolic rates of glucose (LCMRglc) in partial epilepsy,
both in animals and in humans. Patterns of metabolic changes associated
with a wide variety of experimental models of epilepsy have been examined
(Ackerman et al., 1983; Ben Ari et al., 1981; Caveness et al., 1980; Collins
et al., 1976; Engel et al., 1978; Hosokawa et al., 1980; Kato et al., 1980),
which more or less closely resemble the metabolic patterns revealed by
FDG studies in epileptic patients (Bernardi et al., 1983; Engel et al., 1982,
1983; Gur et al., 1982; Kuhl et al., 1980; Theodore et al., 1983). The
common characteristic emerging from animal and human studies of cerebral metabolic changes in partial, convulsive epilepsy is the presence of a
localized anatomical substrate specific to the type of seizure.

By contrast, PET studies in humans have shown that generalized, non-
convulsive seizures are associated with a diffuse increase in LCMRglc having no specific anatomical substrate (Engel et al., 1982, 1985; Ochs et al.,
1987; Theodore et al., 1985). Most animal studies have been performed
with models of petit-mal induced by systemic injection of drugs such as γ-
hydroxybutyrate, GABA agonists or opioids, which produce permanent
spike-and-wave discharges similar to a petit-mal status (Engel et al., 1990;
Theodore, 1988; Chugani et al., 1984a,b; Fariello et al., 1984; Kelly and
McCulloch, 1982; Palacios et al., 1982; Wolfson et al., 1977). The cerebral
metabolic changes recorded vary according to the model used, but no
generalized hypermetabolism has ever been detected; on the contrary, if
anything, even a tendency towards localized hypometabolism has been
observed. These results are, in fact, similar to the results obtained in
humans with atypical absences or petit-mal status (Engel et al., 1990).

Since a model of genetic absence epilepsy has become available using a
strain of Wistar rats selected for spontaneous occurrence of generalized,
nonconvulsive epileptic seizures (Marescaux et al., 1984a,b; Vergnes et al.,
1982) in rats, we decided to measure LCMRglc in these animals by the
[^{14}C]2-deoxyglucose (2DG) quantitative autoradiographic method of
Sokoloff et al. (1977). The electroencephalographic (EEG) and clinical patterns, as well as the pharmacological responses of these seizures, resemble
those of petit-mal epileptic absences in humans (Marescaux et al., 1984b;
Vergnes et al., 1982). Bilateral and synchronous spike-and-wave discharges
are mainly recorded all over the cortex and in the lateral thalamic nuclei,
whereas they do not occur in limbic structures (Vergnes et al., 1987, 1990a).
These spontaneous seizures are suppressed by antiepileptic drugs effective
in the treatment of human petit-mal epilepsy, especially by ethosuximide, a
specific antiabsence drug. By contrast, they are unresponsive to pharmacological agents specific for focal or convulsive seizures (Marescaux et al.,
1984a; Micheletti et al., 1985). This first genetic model of spontaneous,

nonconvulsive seizures in the rat fulfills the requirements for an experimental model of petit-mal epilepsy (Vergnes et al., 1990b). In the present study, LCMRglc was compared in rats from the epileptic strain and their nonepileptic controls. Furthermore, LCMRglc was measured in both strains of rats receiving either haloperidol to induce a permanent petit-mal status or ethosuximide, to suppress the seizures.

Materials and methods

Animals and drug treatment

The experiments were performed on a total of 33 adult male Wistar rats, weighing 310–440 g, randomly chosen from the inbred strains of the laboratory colony and divided into epileptic and control groups of 5 or 6 animals each. These rats were fitted with four single contact electrodes over the cortex, two on each side (Vergnes et al., 1982). Four 30-min EEG traces were recorded from the left and right frontoparietal cortex of all rats to verify the occurrence of spontaneous seizures. On the day of the experiment, a femoral artery and vein were catheterized with polyethylene catheters under light halothane anesthesia. The distal ends of both catheters were passed under the skin to exit through a small incision in the neck, so as to be easily accessible and to allow the animals freedom of movement. The animals were given at least 3 h to recover from anesthesia before measurements of local cerebral glucose utilization were begun. The EEG was recorded throughout the experiment.

In the first experiment, LCMRglc was measured in control and epileptic rats not given any drug treatment. In the second experiment, both controls and epileptic rats received an i.p. injection of 2 mg/kg haloperidol. In the third experiment, both control and epileptic animals received ethosuximide i.p. in a dose of 200 mg/kg. Drugs were administered 10 min before the start of the 2DG procedure.

Physiological functions

Blood pressure in the femoral artery was checked regularly throughout the experiment by means of an air-damped mercury manometer. Just before and 45 min after the administration of 2DG, a sample of arterial blood was taken for measurements of pH, pO_2 and pCO_2 in a blood gas analyzer (Corning, Model 158, Corning Medical and Scientific, Halstead, England). The hematocrit was determined in samples collected in capillary tubes immediately before and 35 min after 2DG injection.

Measurement of local cerebral glucose utilization

Local cerebral glucose utilization was measured by the 2DG method described by Sokoloff et al. (1977). The 2-[^{14}C]deoxyglucose (4.63 MBq/kg; spec. act., 1.65–2.04 GBq/mmol; New England Nuclear, Boston, MA) was injected as an i.v. pulse and timed arterial blood samples were drawn during the following 45 min. The blood samples were immediately centrifuged in a Hettich Microrapid microfuge, and the plasma concentration of 2DG was determined by liquid scintillation counting (Beckman scintillation counter Model LS 1801; Beckman Instruments, Fullerton, CA). Plasma glucose concentrations were measured in a Beckman glucose analyzer. Approximately 45 min after the pulse of 2DG, the animals were killed by decapitation. Brains were rapidly removed and frozen in isopentane chilled to −35°C, coated with chilled embedding medium (carboxymethylcellulose 4% in water), and stored at −80°C in plastic bags until sectioned and autoradiographed. The brains were then cut into 20-μm coronal sections at −22°C in a cryostat. Sections were picked up on glass coverslips and dried on a hot plate (60°C). Sections were autoradiographed on Kodak SB5 film along with [14C]methylmethacrylate standards (Amersham, Arlington Heights, USA) calibrated for their 14C concentration in brain sections, as previously described (Sokoloff et al., 1977). Adjacent sections were fixed and stained with thionin for histological identification of specific nuclei.

The autoradiographs were analyzed by quantitative densitometry with a computerized image-processing system (Biocom 200, France) or a manual microdensitometer (Macbeth, TD901, Kollmorgen Co., Newburgh, USA). Optical density measurements for each structure anatomically defined in Paxinos and Watson's Rat Brain Atlas (1982) were made bilaterally in a minimum of 4 brain sections. All densitometry was performed without knowledge of the epileptic status of the animal. Tissue 14C concentrations were determined from the optical densities of the autoradiographic images of the tissues and a calibration curve obtained from the autoradiographs of the calibrated standards. Glucose utilization was then calculated from the local tissue concentration of 14C, the time-courses of the plasma 2DG and glucose concentrations, and the appropriate constants according to the operational equation of the method (Sokoloff et al., 1977). The lumped constant is expected to change in pathophysiological states such as severe convulsive seizure activity, when the rate of cerebral glucose consumption is increased to the extent that it exceeds the rate of resupply: in fact, during severe convulsive seizure activity, blood glucose concentration falls dramatically, the rate of cerebral glucose utilization becomes limited by the supply, and the lumped constant rises (Sokoloff, 1985). However, in the present study, spike-and-wave discharges did not induce any change in blood glucose levels (Table 1), which allowed us to use the regular lumped constant for calculations.

Statistical analysis

LCMRglc were determined in 16 structures from three control groups and three groups of epileptic rats. Values of LCMRglc in control animals were compared with those in epileptic animals given the same drug, using Student's t-test for independent samples.

Results

Physiological variables

As shown in Table 1, most physiological variables recorded in the present study were similar in both epileptic and control rats and changed only slightly in response to drug treatment. Nevertheless, the strain of spontaneously epileptic rats seemed more sensitive to the drugs than control rats: both haloperidol and ethosuximide induced significant changes vs. control values, mainly in plasma glucose levels and arterial blood gases. Moreover, arterial pH and pCO_2 were already different in epileptic and control rats

Table 1. Physiological variables in control and epileptic rats

	Control (n = 5)	Epileptic (n = 5)	Control + HAL (n = 6)	Epileptic + HAL (n = 6)	Control + ESM (n = 6)	Epileptic + ESM (n = 6)
Plasma glucose concentration (mg/100 ml)						
0 time	189 ± 9	185 ± 12	140 ± 6	164 ± 11[a]	225 ± 10	204 ± 7
10 min after drug injection	183 ± 5	176 ± 13	173 ± 14	216 ± 14[a]*	232 ± 12	199 ± 9[a]
45 min after 2DG administration			171 ± 10	249 ± 11[d]**	189 ± 9	200 ± 17
Arterial blood pressure (mm Hg)						
0 time	121 ± 4	118 ± 3	123 ± 2	127 ± 1	123 ± 1	121 ± 2
10 min after drug injection	111 ± 3	116 ± 4	113 ± 2	112 ± 2	113 ± 2	115 ± 3
35 min after 2DG administration			104 ± 5	86 ± 6[a]*	110 ± 1	115 ± 3
Arterial pH						
0 time	7.42 ± 0.01	7.45 ± 0.01[a]	7.47 ± 0.01	7.45 ± 0.01	7.44 ± 0.01	7.42 ± 0.01
10 min after drug injection	7.44 ± 0.01	7.44 ± 0.01	7.46 ± 0.01	7.46 ± 0.01	7.45 ± 0.01	7.43 ± 0.01
45 min after 2DG administration			7.47 ± 0.01	7.44 ± 0.01[a]	7.44 ± 0.01	7.42 ± 0.01
Arterial pO_2 (mm Hg)						
0 time	83 ± 3	90 ± 4	81 ± 2	86 ± 3	80 ± 3	84 ± 3
10 min after drug injection	86 ± 4	90 ± 2	78 ± 5	91 ± 3[a]	86 ± 3	91 ± 1[a]
45 min after 2DG administration			78 ± 3	93 ± 4[b]	85 ± 2	92 ± 2[a]*
Arterial pCO_2 (mm Hg)						
0 time	41 ± 1	35 ± 1[a]	37 ± 1	38 ± 2	41 ± 2	40 ± 2
10 min after drug injection	40 ± 1	37 ± 1	39 ± 2	35 ± 1[a]	39 ± 1	37 ± 1
45 min after 2DG administration			37 ± 1	35 ± 1	41 ± 2	39 ± 1
Hematocrit (%)	44 ± 3	42 ± 2	44 ± 1	42 ± 1	44 ± 1	45 ± 1

Values are means ± S.E.M. of the number of animals in parentheses.
[a]$P < 0.05$, [b]$P < 0.01$, [d]$P < 0.0005$, statistically significant difference from the corresponding control.
*$P < 0.05$, **$P < 0.01$, statistically significant difference from zero time

that had received no drug treatment at the beginning of the 2DG procedure. However, all physiological parameters remained within the normal physiological range and were not likely to have had any significant influence on the levels of cerebral energy metabolism measured in the present study in both control and epileptic animals.

Electroencephalographic activity

EEG tracings of epileptic rats showed bilateral, synchronous spike-and-wave discharges of 7–10 c/s and 250–800 μV amplitude, i.e. exceeding at least five times above the amplitude of the background desynchronized EEG. The number of spike-and-wave discharges over the 45 min of the 2DG experiment in the 5 epileptic rats not given any drug was 61 ± 10 (mean $\pm$ S.E.M.). Their duration ranged from 3 to 45 sec with a mean length of 16.6 ± 1.7 sec per min of recording. In epileptic rats treated with haloperidol, the mean length of spike and wave discharges over the 45 min of the 2DG experiment was 44.2 ± 4.9 sec per min of recording. By contrast, ethosuximide totally abolished spike-and-wave discharges throughout the experimental period. The EEG patterns of control animals, drug-treated or untreated were normal, with fast desynchronized activity, confirming that they were awake during the experiment. The administration of either haloperidol or ethosuximide induced no changes in EEG recordings or in the general behavior of control rats. The animals were usually calm and showed occasional spontaneous locomotion. 2DG was injected only when

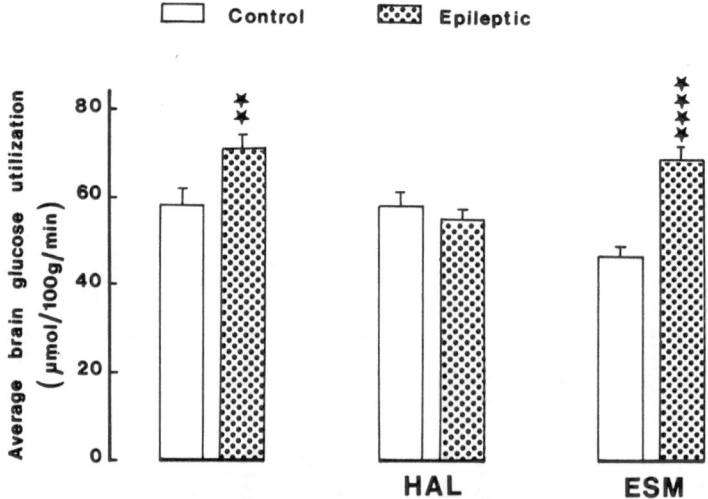

Fig. 1. Average rates of brain glucose utilization, expressed as μmol/100 g/min $\pm$ S.E.M., in control and epileptic rats, not exposed to any drug (left bars), injected with haloperidol (HAL, middle bars), or ethosuximide (ESM, right bars). **P < 0.01, ****P < 0.0005, statistically significant difference from the corresponding control

spike-and-wave discharges were occurring at a constant rate and/or the animals were in a calm state and induced neither a blokade of spike-and-wave discharges nor any change in the length of absence episodes or the general behavior of the rats.

Average brain glucose utilization

Average brain glucose utilization, as obtained from 60 brain structures reached a value of 58 µmol/100 g/min in both controls given no drug treatment and controls receiving haloperidol (Fig. 1). Administration of ethosuximide to control rats, on the other hand, led to a 19% decrease in mean cerebral glucose utilization, as compared to untreated controls. In untreated epileptic rats and those given ethosuximide, average brain glucose utilization was increased over control values by 22 and 47%, respectively. By contrast, there was no difference in the mean value of cerebral glucose utilization between controls and epileptic rats given haloperidol (Fig. 1).

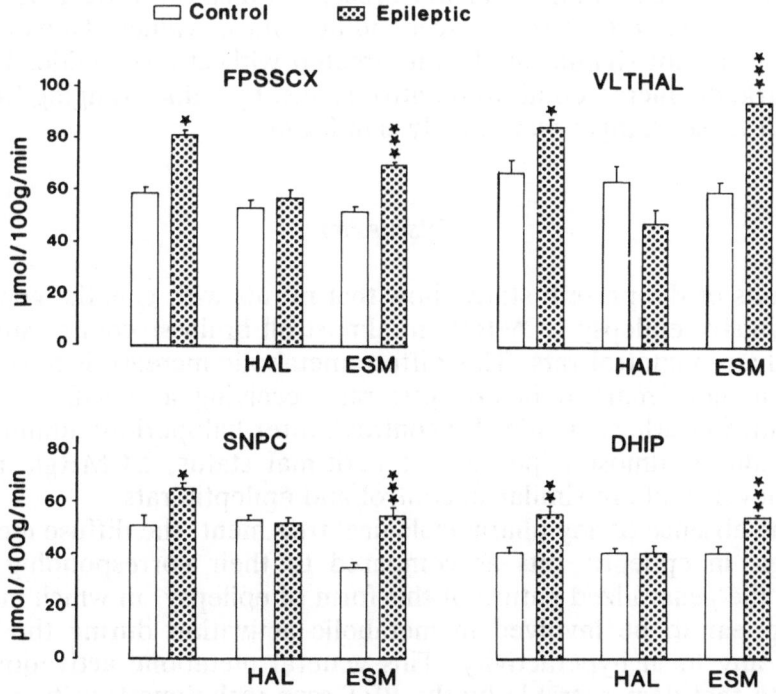

Fig. 2. Rates of glucose utilization, expressed as µmol/100 g/min ± S.E.M., in different areas of the brain of control and epileptic rats not exposed to any drug (left bars), injected with haloperidol (HAL, middle bars), or ethosuximide (ESM, right bars). Abbreviations: *FPASS* frontoparietal cortex, somatosensory area; *VLTHAL* ventrolateral thalamus; *SNPC* substantia nigra, pars compacta; *DHIP* dorsal hippocampus. *P < 0.05, **P < 0.01; ***P < 0.005, statistically significant difference from the corresponding control

Local cerebral glucose utilization

In four areas representative of different systems of the brain, similar levels of LCMRglc were found in untreated control rats and those receiving haloperidol (Fig. 2). In control rats given ethosuximide, LCMRglc was the same as in normal controls in the somatosensory area of the frontoparietal cortex and dorsal hippocampus, but lower in the ventrolateral thalamus and substantia nigra pars compacta. In epileptic rats, LCMRglc was above the corresponding control values in both the untreated and ethosuximide-treated animals. When haloperidol was given, there was no longer any difference in LCMRglc between epileptic rats and the corresponding controls (Fig. 2).

The percentage of variation of LCMRglc in epileptic rats compared to the corresponding controls is shown in Fig. 3 for 14 brain areas belonging to the cortex, thalamus, motor and limbic systems, and nigrotectal pathway. LCMRglc in untreated epileptic rats was increased above control values by 10 to 42%; the changes were significant in 12 of the 14 structures studied. In haloperidol-treated epileptic rats, LCMRglc differed from control values by −17 to +8%, except in the thalamic areas, in which there were decreases of 16 to 25% compared to the corresponding control values. However, these changes were not significant. In rats treated with ethosuximide, LCMRglc were markedly increased above control levels, by values ranging from 31 to 72%. All these changes were highly significant.

Discussion

The results of the present study show that in rats with genetic, generalized, nonconvulsive epilepsy LCMRglc in almost all brain structures studied are higher than in control rats. This diffuse metabolic increase is also apparent and even more marked in epileptic rats receiving a specific antiabsence treatment, i.e. ethosuximide. By contrast, after haloperidol administration, which induces almost a permanent petit-mal status, LCMRglc revert to normal levels and are similar in control and epileptic rats.

In the absence of any pharmacological treatment, the diffuse increases in LCMRglc in epileptic rats as compared to their corresponding controls confirm the generalized nature of this form of epilepsy, in which most brain areas appear to be involved in metabolic activation during the seizures, without any focal hyperactivity. This general metabolic activation is very similar to that demonstrable by the PET scan technique (positron emission tomography) in patients with typical childhood absence epilepsy, whose cerebral metabolism is massively increased when absence seizures occur during the scan, with no selective substrate for the seizures (Engel et al., 1982, 1985, 1988, 1990). It then appears from human and animal studies that there is no anatomical correlation between the areas in which LCMRglc are increased and those generating EEG spike-and-wave discharges. In

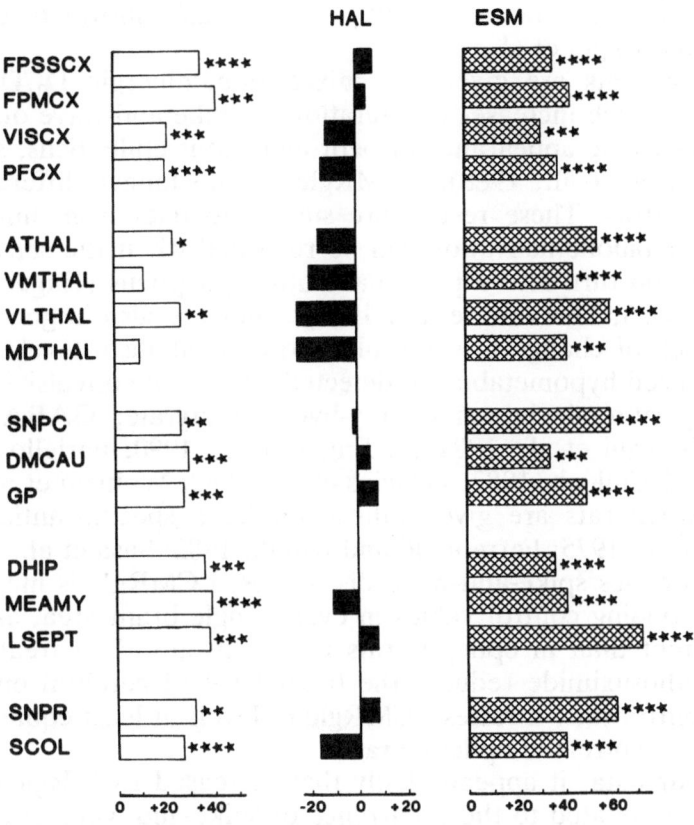

Fig. 3. Rates of glucose utilization in cerebral areas of epileptic rats, expressed as percent of variation from the corresponding control, in rats not exposed to any drug (left column), injected with haloperidol (HAL, middle column), or with ethosuximide (ESM, right column). Abbreviations: *FPASS* frontoparietal cortex, somatosensory area; *FPAM* frontoparietal cortex, motor area; *VISCX* visual cortex; *PFTALCX* prefontal cortex; *ATHAL* anterior thalamus; *VMTHAL* ventromedian thalamus; *VLTHAL* ventrolateral thalamus; *MDTHAL* mediodorsal thalamus; *SNPC* substantia nigra, pars compacta; *DMCAU* dorsomedial caudate nucleus; *GP* globus pallidus; *DHIP* dorsal hippocampus; *MEAMY* medial amygdala; *LSEPT* lateral septum; *SNPR* substantia nigra, pars reticulata; *SCOL* superior colliculus. *P < 0.05, **P < 0.01, ***P < 0.005, ****P < 0.0005, statistically significant difference from the corresponding control

fact, previous recordings of cortical and subcortical bipolar EEG activity throughout the brains of rats exhibiting genetic absence epilepsy have shown that spike-and-wave discharges are most prominent in the neocortex and thalamus, but do not appear in the limbic system (Vergnes et al., 1987, 1990a). By contrast, in the present study, LCMRglc is increased in almost all brain structures in epileptic rats, whether or not they exhibit spike-and-wave discharges, demonstrating that LCMRglc and EEG activity do not reflect the same functional activity, as previously shown in other studies (Engel, 1988; Engel et al., 1990).

When the rats are given the mixed dopaminergic D_1/D_2 antagonist haloperidol, which increases the duration of spike-and-wave discharges and almost induces the appearance of petit-mal status epilepticus, as previously shown (Warter et al., 1988), LCMRglc are no longer different from the baseline controls. These results are similar to data from human studies, where no metabolic activation can be recorded when the seizures become permanent and turn into a petit-mal status epilepticus (Engel et al., 1990; Theodore, 1988; Theodore et al., 1985). They are also in good accordance with the lack of change in cerebral energy metabolism levels, or with the rather localized hypometabolism detected when non-convulsive seizures are induced by systemic injection of γ-hydroxybutyrate, GABA agonists or opioids (Chugani et al., 1984a,b; Engel et al., 1990; Fariello et al., 1984; Kelly and McCulloch, 1982; Palacios et al., 1982; Wolfson et al., 1987). By contrast, when rats are given ethosuximide, a specific antiabsence drug (Browne et al., 1975; Ferrendelli and Klunk, 1982; Sasa et al., 1988), which totally suppresses spike-and-wave discharges, LCMRglc is increased above the corresponding control values in every single brain area, and even to a greater extent than in epileptic rats not given any drug treatment. Thus, although ethosuximide reduces the basal level of cerebral energy metabolism in control rats, it raises LCMRglc to levels at least equal to or higher than those in untreated epileptic rats.

From our data, it appears likely that increased LCMRglc is not, or at least not only related to the occurrence of spike-and-wave discharges. This lack of correlation between the anatomical localization of spike-and-wave discharges and the changes in cerebral energy metabolism levels in epileptic rats is further confirmed by the data in epileptic animals treated with pharmacological agents, which show a mainly diffuse and fairly large increase over baseline control values of LCMRglc after ethosuximide treatment and no change in response to haloperidol. It then appears likely that, rather than occurrence of spike-and-wave discharges, processes involved in their termination and suppression as well as in the prevention of their spread to highly susceptible limbic structures in rats with spontaneous petit-mal-like seizures may represent energy consuming activities. Moreover, absence seizures have no form of motor expression, except for rare facial twitches, whereas spike-and-wave discharges involve cortical motor areas and their subcortical connections (Vergnes et al., 1990a). So, whatever the nature of the mechanism involved in the blockade of motor expression was, it may contribute to the increase in metabolic activity, since, as previously shown, inhibitory as well as excitatory activities are correlated with increased metabolic activity (Ackerman et al., 1984). Furthermore, the marked hypermetabolism recorded in typical childhood epilepsy has been rather attributed to prominent inhibitory mechanisms (Engel et al., 1990).

In conclusion, the diffuse increase in cerebral energy metabolism in rats with spontaneous, generalized, nonconvulsive epilepsy is similar to that observed in humans with typical genetic absences. This increased metabolism does not appear to be directly correlated either with the number and duration of the spike-and-wave discharges, or with their anatomical distribution. As confirmed by the data on the effects of ethosuximide, the increased cerebral metabolism of epileptic rats may rather be the result of inhibitory mechanisms preventing the occurrence and spread of spike-and-wave discharges.

Acknowledgements

This work was supported by a grant from INSERM (CAR n° 4900019).

References

Ackermann RF, Finch DM, Babb TL, Engel J Jr (1984) Increased glucose metabolism during long-duration recurrent inhibition of hippocampal pyramidal cells. J Neurosci 4: 251–264

Ben-Ari Y, Tremblay E, Riche D, Ghilini G, Naquet R (1981) Electrographic, clinical pathological alterations following systemic administration of kainic acid, bicuculline or pentetrazole: metabolic mapping using the deoxyglucose method with special reference to the pathology of epilepsy. Neuroscience 6: 1361–1391

Bernardi S, Trimble MR, Frackowiak RSJ, Wise RJS, Jones T (1983) An interictal study of partial epilepsy using positron emission tomography and the oxygen-15 inhalation technique. J Neurol Neurosurg Psychiatry 46: 473–477

Browne TR, Dreifuss FE, Dyken PR, Goode DJ, Penry JK, Porter RJ, White BG, White PT (1975) Ethosuximide in the treatment of absence (petit mal) seizures. Neurology 25: 515–524

Caveness WF, Kato M, Malamut BL, Hosokawa S, Wakisaka S, O'Neill RR (1980) Propagation of focal motor seizures in the pubescent monkey. Ann Neurol 7: 213–221

Chugani HT, Ackermann RF, Chugani DC, Engel J Jr (1984) Autoradiographic studies of opioid-mediated epileptogenic phenomena in rats. In: Fariello RG, Morselli PL, Lloyd KG, Quesney LF, Engel J Jr (eds) Neurotransmitters, seizures and epilepsy II. Raven Press, New York, pp 315–325

Chugani HT, Ackermann RF, Chugani DC, Engel J Jr (1984) Opioid-induced epileptogenic phenomena: anatomical, behavioral, and electroencephalographic features. Ann Neurol 15: 361–368

Collins RC, Kennedy C, Sokoloff L, Plum F (1976) Metabolic anatomy of focal motor seizures. Arch Neurol 33: 536–542

Engel J Jr (1988) Comparison of positron emission tomography and electroencephalography as measures of cerebral function in epilepsy. In: Pfurtscheller G, Lopes da Silva FH (eds) Functional brain imaging. Hans Huber, Bern, pp 229–238

Engel J Jr, Ludwig BI, Fetell M (1978) Prolonged partial complex status epilepticus: EEG and behavioral observations. Neurology 28: 863–869

Engel J Jr, Kuhl DE, Phelphs ME (1982) Patterns of human local cerebral glucose metabolism during epileptic seizures. Science 218: 64–66

Engel J Jr, Kuhl DE, Phelps ME, Rausch R, Nuwer M (1983) Local cerebral metabolism during partial seizures. Neurology 33: 400–413

Engel J Jr, Lubens P, Kuhl DE (1985) Local cerebral metabolic rate for glucose during petit mal absences. Ann Neurol 17: 121–128

Engel J Jr, Lubens P, Phelps M (1988) Metabolic correlates of diffuse EEG spike-and-wave and absence seizures. Ann Neurol 23: 207–208

Engel J Jr, Ochs R, Gloor P (1990) Metabolic studies of generalized epilepsy. In: Avoli M, Gloor P, Kostopoulos G, Naquet R (eds) Generalized epilepsy: neurobiological approaches. Birkhäuser, Boston, pp 387–396

Fariello RG, Golden GT, Reyes PF (1984) Metabolic correlates of GABAmimetic-induced EEG abnormalities. In: Fariello RG, Morselli PL, Lloyd KG, Quesney LF, Engel J Jr (eds) Neurotransmitters, seizures and epilepsy II. Raven Press, New York, pp 245–252

Ferrendelli JA, Klunk WE (1982) Ethosuximide. Mechanisms of action. In: Woodbury DM, Penry JK, Pippenger CE (eds) Antiepileptic drugs. Raven Press, New York, pp 655–661

Gur RC, Sussman NM, Alavi A, Gur RE, Rosen AD, O'Connor M, Goldberg HI, Greenberg JH, Reivich M (1982) Positron emission tomography in two cases of childhood epileptic encephalopathy (Lennox-Gastaut syndrome). Neurology 32: 1191–1194

Hosokawa S, Iguchi T, Caveness WF, Kato M, O'Neill RR, Wakisaka S, Malamut BL (1980) Effects of manipulation of the sensorimotor system on focal motor seizures in the monkey. Ann Neurol 7: 222–229

Kato M, Malamut BL, Caveness, WF, Hosokawa S, Wakisaka S, O'Neill RR (1980) Local cerebral glucose utilization in newborn and pubescent monkeys during focal motor seizures. Ann Neurol 7: 204–212

Kelly PAT, McCulloch J (1982) Effects of the putative GABAergic agonists, muscimol and THIP, upon local cerebral glucose utilization. J Neurochem 39: 613–624

Kuhl DE, Engel J Jr, Phelps ME, Selin C (1980) Epileptic patterns of local cerebral metabolism and perfusion in humans determined by emission computed tomography of ^{18}FDG and $^{13}NH_3$. Ann Neurol 8: 348–360

Marescaux C, Micheletti G, Vergnes M, Depaulis A, Rumbach L, Warter JM (1984a) A model of chronic spontaneous petit mal-like seizures in the rat: comparison with pentylenetetrazol-induced seizures. Epilepsia 25: 326–331

Marescaux C, Vergnes M, Micheletti G, Depaulis A, Rumbach L, Warter JM, Kurtz D (1984b) Une forme génétique d'absences petit-mal chez le rat Wistar. Rev Neurol 140: 63–66

Micheletti G, Vergnes M, Marescaux C, Reis J, Depaulis A, Rumbach L, Warter JM (1985) Antiepileptic drug evaluation in a new animal model: spontaneous petit mal epilepsy in the rat. Arzneimittelforschung/Drug Res 35: 483–485

Ochs RF, Gloor P, Tyler JL, Wolfson T, Worsley K, Anderman F, Diksic M, Meyer E, Evans A (1987) Effect of generalized spike-and-wave discharge on glucose metabolism measured by positron emission tomography. Ann Neurol 21: 458–464

Palacios JM, Kuhar MJ, Rapoport SI, London ED (1982) Effects of γ-aminobutyric acid agonist and antagonist drugs on local cerebral glucose utilization. J Neurosci 2: 853–860

Paxinos G, Watson C (1982) The rat brain in stereotaxic coordinates. Academic Press, New York

Phelps ME, Huang SC, Hoffman EJ, Selin C, Sokoloff L, Kuhl DE (1979) Tomographic measurement of local cerebral glucose metabolic rate in humans with (F-18)2-fluoro-2-deoxy-d-glucose: validation of method. Ann Neurol 6: 371–388

Reivich M, Kuhl D, Wolf A, Greenberg J, Phelps M, Ido T, Cassella V, Fowler J, Hoffman E, Alavi A, Som P, Sokoloff L (1979) The [^{18}F]fluoro-deoxyglucose method for the measurement of local cerebral glucose utilization in man. Circ Res 44: 127–137

Sasa M, Ohno Y, Ujihara H, Fujita Y, Yoshimura M, Takaori S, Serikawa T, Yamada J (1988) Effects of antiepileptic drugs on absence-like and tonic seizures in the spontaneously epileptic rat, a double mutant rat. Epilepsia 29: 505–513

Sokoloff L (1985) Basic principles in imaging of regional cerebral metabolic rates. In: Sokoloff L (ed) Brain imaging and function. Raven Press, New York, pp 21–49

Sokoloff L, Reivich M, Kennedy C, Des Rosiers MH, Patlak CS, Pettigrew K, Sakurada O, Shinohara M (1977) The [^{14}C]deoxyglucose method for the measurement of local cerebral glucose utilization: theory, procedure and normal values in the conscious and anesthetized albino rat. J Neurochem 28: 897–916

Theodore WH (1988) Cerebral metabolism in absence seizures and related syndromes. In: Myslobodsky MS, Mirsky AF (eds) Elements of petit mal epilepsy. Peter Lang, New York, pp 131–157

Theodore WH, Newmark ME, Sato S, Brooks R, Patronas M, De La Paz R, DiChiro G, Kessler RM, Margolin R, Manning RG, Channing M, Porter R (1983) [^{18}F]Fluorodeoxyglucose positron emission tomography in refractory complex partial seizures. Ann Neurol 14: 429–437

Theodore WH, Brooks R, Margolin R, Patronas N, Sato S, Porter RJ, Mansi L, Bairamian D, DiChiro G (1985) Positron emission tomography in generalized seizures. Neurology 35: 684–690

Vergnes M, Marescaux C, Micheletti G, Reis J, Depaulis A, Rumbach L, Warter JM (1982) Spontaneous paroxysmal electroclinical patterns in rat: a model of generalized non-convulsive epilepsy. Neurosci Lett 33: 97–101

Vergnes M, Marescaux C, Depaulis A, Micheletti G, Warter JM (1987) Spontaneous spike and wave discharges in thalamus and cortex in a rat model of genetic petit mal-like seizures. Exp Neurol 96: 127–136

Vergnes M, Marescaux C, Depaulis A (1990a) Mapping of spontaneous spike and wave discharges in Wistar rats with genetic generalized non convulsive epilepsy. Brain Res 523: 87–91

Vergnes M, Marescaux C, Depaulis A (1990b) The spontaneous spike and wave discharges in Wistar rats: a model of genetic generalized non convulsive epilepsy. In: Avoli M, Gloor P, Kostopoulos G, Naquet R (eds) Generalized epilepsy: neurobiological approaches. Birkhäuser, Boston, pp 238–253

Warter JM, Vergnes M, Depaulis A, Tranchant C, Rumbach L, Micheletti G, Marescaux C (1988) Effects of drugs affecting dopaminergic neurotransmission in rats with spontaneous petit mal-like seizures. Neuropharmacology 27: 269–274

Wolfson LI, Sakurada O, Sokoloff L (1977) Effects of γ-butyrolactone on local cerebral glucose utilization in the rat. J Neurochem 29: 777–783

Authors' address: Dr. A. Nehlig, INSERM U272, Université de Nancy I, 30, rue Lionnois, B.P. 3069, F-54013 Nancy Cedex, France

J Neural Transm (1992) [Suppl] 35: 155–177

Experimental absence seizures: potential role of γ-hydroxybutyric acid and GABA$_B$ receptors

R. Bernasconi[1], J. Lauber[1], C. Marescaux[2], M. Vergnes[3], P. Martin[1], V. Rubio[1], T. Leonhardt[1], N. Reymann[1], and H. Bittiger[1]

[1] Research and Development Department, Pharmaceuticals Division, Ciba-Geigy, Basel, Switzerland
[2] Groupe de Recherche de Physiologie Nerveuse, Clinique Neurologique, Hospices Civils, and [3] Centre de Neurochimie du CNRS et de l'INSERM, Strasbourg, France

Summary. We have investigated whether the pathogenesis of spontaneous generalized non-convulsive seizures in rats with genetic absence epilepsy is due to an increase in the brain levels of γ-hydroxybutyric acid (GHB) or in the rate of its synthesis. Concentrations of GHB or of its precursor γ-butyrolactone (GBL) were measured with a new GC/MS technique which allows the simultaneous assessment of GHB and GBL. The rate of GHB synthesis was estimated from the increase in GHB levels after inhibition of its catabolism with valproate. The results of this study do not indicate significant differences in GHB or GBL levels, or in their rates of synthesis in rats showing spike-and-wave discharges (SWD) as compared to rats without SWD. Binding data indicate that GHB, but not GBL, has a selective, although weak affinity for GABA$_B$ receptors ($IC_{50} = 150\,\mu M$). Similar IC_{50} values were observed in membranes prepared from rats showing SWD and from control rats. The average GHB brain levels of $2.12 \pm 0.23\,nmol/g$ measured in the cortex and of $4.28 \pm 0.90\,nmol/g$ in the thalamus are much lower than the concentrations necessary to occupy a major part of the GABA$_B$ receptors. It is unlikely that local accumulations of GHB reach concentrations 30–70-fold higher than the average brain levels. After injection of $3.5\,mmol/kg$ GBL, a dose sufficient to induce SWD, brain concentrations reach $240 \pm 31\,nmol/g$ (Snead, 1991) and GHB could thus stimulate the GABA$_B$ receptor.

Like the selective and potent GABA$_B$ receptor agonist R(−)-baclofen, GHB causes a dose-related decrease in cerebellar cGMP. This decrease and the increase in SWD caused by R(−)-baclofen were completely blocked by the selective and potent GABA$_B$ receptor antagonist CGP 35348, whereas only the increase in the duration of SWD induced by GHB was totally antagonized by CGP 35348. The decrease in cerebellar cGMP levels elicited by GHB was only partially antagonized by CGP 35348.

These findings suggest that all effects of R(−)-baclofen are mediated by the $GABA_B$ receptor, whereas only the induction of SWD by GHB is dependent on $GABA_B$ receptor mediation, the decrease in cGMP being only partially so. Taken together with the observations of Marescaux et al. (1992), these results indicate that $GABA_B$ receptors are of primary importance in experimental absence epilepsy and that $GABA_B$ receptor antagonists may represent a new class of anti-absence drugs.

1. Introduction

Primary generalized epilepsy of the absence type is a childhood-onset seizure disorder of unknown etiology characterized behaviourally by brief staring spells and arrest of motor activity, and electrically by generalized 3 Hz spike-and-wave discharges (SWD) in the electroencephalogram (EEG) (Godschalk et al., 1976, 1977; Mirsky et al., 1986). Three Hz SWD are associated with enhanced GABA-mediated synaptic inhibition and absence epilepsy could conceivably represent generalized inhibitory seizures due to an excess, rather than to a deficit of GABA-mediated transmission (Fariello and Golden, 1987; Fromm and Kohli, 1972; Gloor and Fariello, 1988). Evidence for this premise is based on the fact that direct $GABA_A$ and $GABA_B$ receptor agonists, GABA uptake inhibitors and 4-aminobutyrate: 2-oxoglutarate aminotransferase (EC 2.6.1.19; GABA-T) inhibitors augment the number and duration of discharges (King, 1979; Marescaux et al., 1984; Micheletti et al., 1985; Smith and Bierkamper, 1990; Snead, 1990; Vergnes et al., 1984). The GABA metabolite and/or putative neurotransmitter (Vayer et al., 1987) γ-hydroxybutyric acid (GHB) or its lactonized prodrug γ-butyrolactone (GBL), also induces 4–6 Hz SWD accompanied by arrest of motor activity, with staring, facial myoclonus and vibrissal twitches, which mimic the events of absence seizures in rats (Snead et al., 1976; Snead, 1988). As the changes in EEG observed after administration of GHB are not followed by convulsions, GHB-induced seizures have been proposed as an animal model of petit mal epilepsy (Godschalk et al., 1976, 1977).

Because of the structural resemblance of GHB to GABA, GHB has also been described as a "GABA agonist" (Meldrum, 1981), suggesting that the epileptiform discharges caused by GHB may be due to its GABAergic activity. In agreement with this hypothesis, Pericic et al. (1978) have shown that GHB, like $GABA_A$ agonists, does not alter GABA levels, but produces a marked and dose-related reduction in the rate of GABA synthesis, indicating strong interactions between GHB and GABA-mediated inhibition. In contrast to the action of muscimol, this effect is not secondary to a direct effect of GHB on $GABA_A$ receptors (Enna and Snyder, 1975). Thus, GHB modulates GABA neurotransmission and induces absence-like seizures by way of a mechanism which is not mediated through $GABA_A$ receptors.

Since exogeneous GHB is capable of inducing absence seizures, the question naturally arises whether GHB-mediated mechanisms might play a role in the genesis of petit mal epilepsy. One possibility of testing the GHB hypothesis of petit mal epilepsy is to assess biochemical parameters related to GHB activity in the brain (e.g. GHB levels, its rate of synthesis, GHB binding or second messengers) in animals with absence seizures as compared to non-epileptic animals.

Recently, a genetic model of spontaneous generalized non-convulsive seizures has been described (Vergnes et al., 1982), which satisfies most of the criteria proposed for a useful animal model of petit mal epilepsy (Mirsky et al., 1986). Spontaneous and recurrent SWD were originally seen in the EEG of some Wistar rats (Vergnes et al., 1982). By successive inbreeding of such rats, a strain in which spontaneous SWD can be recorded in 100% of the animals has been selected and named the Genetic Absence Epilepsy Rats from Strasbourg (GAERS) (Vergnes et al., 1987). Concurrently, another strain of rats was selected which never displayed SWD (controls). Both the electrographic characteristics and pharmacological response of these SWD are reminiscent of petit mal epilepsy in man (Vergnes et al., 1982; Micheletti et al., 1985). The GAERS strain thus affords a reproducible and pharmacologically specific model for the study of biochemical mechanisms involved in spontaneous generalized non-convulsive seizures (Engel et al., 1990).

The aim of the present study was to examine the involvement of GHB and GBL in such seizures by measuring the endogenous concentration of both in hippocampus, thalamus and frontal cortex in GAERS and to compare them with the levels in seizure-free rats. The increase in GHB induced by valproate, an index of its rate of synthesis, was also examined in both strains. Levels of GHB and GBL were assessed by a new capillary gas chromatography-mass spectrometry method with selected-ion monitoring (GC-MS) which allows simultaneous measurement of GHB and GBL with the necessary sensitivity. As the SWD induced by GHB are antagonized by the selective $GABA_B$ receptor antagonist CGP 35348 (Marescaux et al., 1992), the interactions of GHB and of its prodrug GBL with 12 neurotransmitter receptors and neuromodulator binding sites, in particular those controlling GABA-mediated inhibition, were evaluated. In addition, we assessed the effects of GBL and of the selective $GABA_B$ receptor agonist R(−)-baclofen either alone or in combination with CGP 35348, on cGMP levels and we used this paradigm to study the potential interactions between GHB and $GABA_B$ receptors in vivo.

2. Material and methods

Animals

Experiments for the development of the GC-MS procedure and for the assessment of cGMP were conducted on male Tif: RAIF (SPF) rats (Tierfarm Sisseln, Switzerland) weighing 240–280 g. Other experiments on cGMP levels were performed on male Tif: MAGf (SPF) mice, 23–27 g body weight, 5–8 weeks of age (Tierfarm Sisseln, Switzerland). The animals were kept in an air-conditioned room at 21°C, with a 12 hour light-dark cycle and were sacrificed between 8:30 and 10:00 a.m. to avoid circadian variations of the different biochemical parameters measured.

Rats with spontaneous absence-like seizures

Male Wistar rats (350–400 g) from the breeding colony at the Centre de Neurochimie, C.N.R.S., Strasbourg were used in this study. They were chosen from the 9th generation of a strain with spontaneous generalized non-convulsive seizures, in which bilateral SWD (frequency = 7–9 c/sec, amplitude = 300–1,000 μV, mean duration = 6.0 ± 3.4 sec with a variance between 0.5 and 40 sec, occurrence = 1/min) are observed in awake but inactive animals. Controls were also from the 9th generation of a strain which never displayed SWD. Epileptic and non-epileptic rats were of the same age. They were sacrificed after an acclimatization period of 15 days in Basel.

Chemicals for GC and GC-MS

1,3-Diphenyl-1,1,3,3-tetramethyldisilazane (DPTMDS, Cat. No 43340), hexamethylsilazane (HMDS, Cat No 52619), acetonitrile, and acetic acid anhydride were purchased from Fluka. The internal standard for GHB and GBL, GBL-2,2,3,3,4,4-d_6 (GBL-d_6) was from Merck, Sharp & Dohme Ltd, Pointe Claire, Quebec, Canada. The stationary phase CP-51 wax was from Chrompack International (Middleburg, The Netherlands). All other chemicals and reagents were of analytical-reagent grade and were used without purification.

Drugs

GHB (sodium salt) and GBL were purchased from Fluka. [2,3-^{3}H]GHB, potassium salt (spec. act. 100 Ci/mmol) was prepared by the CEA (Gif-sur-Yvette, France). Valproate sodium was synthesized in our laboratories by Dr. H. Allgeier. Drugs were dissolved in saline 0.9% such that the volume of injection was 1 ml/kg and were used on the same day, if necessary the pH was adjusted to pH = 5 with NaOH 1N. We only used subanaesthetic (200–400 mg/kg) doses of GBL, which produce EEG and behavioural changes corresponding to stage 1 and 2 of Snead (1988). Doses larger than 400 mg/kg i.p. are associated with a burst suppression pattern described as stage 3 by Snead (1988).

Sample preparation for GC-MS-analysis

Rats were killed by fast focused microwave irradiation of the head (Püschner GmbH, Schwanewede, F.R.G; 1.6 sec, 7.5 kW). The brains were rapidly removed, cooled on dry ice and dissected immediately into different brain areas according to the method of Glowinski and Iversen (1966). The brain structures were divided into two equal parts (left and right). One part of the samples was homogenized for 10 min at room temperature in a ground-glass homogenizer with 2 ml acetonirile containing 20 ng of the the internal standard GBL-d_6. Since GHB does not undergo lactonisation under these conditions (Vayer et al., 1988; Snead et al., 1989), any GHB present would not be lactonized and thus not extracted into the acetonitrile. Therefore, the values obtained represent only GBL. The contralateral brain structures were extracted for 10 min at room temperature in the ground-glass homogenizer with a solution of 5% acetic anhydride in acetonitrile containing 80 ng of GBL-d_6. This procedure lactonizes all the GHB present in the sample, such that the value obtained represents GHB plus GBL. Hence, by subtracting the value obtained from the pure acetonitrile extract, it is possible to determine the concentration of GHB. The acetonitrile solutions were allowed to stand for 1 hr at room temperature and were centrifuged at 10,000 g for 1 hr at 4°C. Owing to the selectivity of the GC-MS method, prior purification of these solutions of GBL in acetonitrile or acetic anhydride/acetonitrile is not necessary.

GC-MS assay for GHB and GBL

The GC-MS analyses were carried out on a Finnigan 4500 mass spectrometer interfaced with an Incos data-processing system and coupled to a Carlo Erba gas chromatograph model 5160, Mega series equipped with the Ciba-Geigy injector model 1988 and the A 200S autosampler. The injector developed at Ciba-Geigy (Lauber-Injector) can be variously operated for split/splitless injection mode, for cold quasi on column injection mode or for hot quasi on column injection mode. All three injection techniques were automated by the autosampler A 200S from Carlo Erba Instruments, Milan, Italy for Europe and Leap Technology, Chapel Hill, NC, for USA.

For GBL analysis, the temperature-controlled "cold on column" mode was chosen. The temperature was kept at 20°C. GC analyses were performed with a 50 m × 0.3 mm glass capillary column pretreated and coated with CP 51 Wax at a film thickness of 1 μm according to Grob (1986), with a 25 m retention gap. The GC oven programme started at 70°C, increased at a rate of 7.5°C per minute to 220°C and was kept for 10 min at this temperature. The temperature of the GC-MS interface and the ion source were kept constant at 250°C and 100°C, respectively. Hydrogen was used as carrier gas at a pressure of 80 kPa. The mass spectra were obtained in the total ion current (TIC) mode. The following mass spectrometric conditions were used: positive chemical ionization with methane as reactant gas at an ion source pressure of 45 kPa measured with an uncalibrated thermocouple gauge. The filament current was kept at 200 μA, and the electron energy at 70 eV. The mass spectrometer was scanned from m/z 50 to 250 daltons in 1 sec intervals. Multiple Ion Detection was used for sensitive, selective simultaneous mass specific detection of the GBL-d_0 and GBL-d_6 at m/z 87 and 93. These base peaks were used for the quantitative assessment of GHB and GBL in brain structures (Fig. 1B). Under these conditions the retention time for GBL and the internal standard GBL-d_6 was 8:15 min (Fig. 1A). Every sample was injected twice.

Measurement of GHB rate of synthesis

The time-dependent accumulation of GHB and GBL following a dose of 400 mg/kg valproate was determined from 0 to 240 min at fixed intervals. GHB and GBL levels were determined by the GC-MS method previously described. Turnover rates were estimated by measuring the accumulation of GHB and GBL in the linear part of the curves obtained.

Receptor binding assays

To demonstrate the selectivity of the interactions, GHB and GBL were tested in a battery of 12 assays including $GABA_A$, $GABA_B$, benzodiazepine, α_1, α_2 and β-adrenoceptors, muscarinic cholinergic, $5HT_1$, histamine H1, adenosine A1, opiate μ and substance P receptors. Methods for receptor-binding assays used in the present investigation are documented in table 3. All assays were validated using appropriate reference standards. When testing the affinity of GHB for $GABA_B$ receptors in epileptic as compared to non-epileptic conrol rats, we used the potent and selective tritiated $GABA_B$ receptor agonist 3-aminopropylphosphinic acid, [^{3}H]CGP 27492 (15.0 Ci/mmol, Ciba-Geigy Horsham, UK) as described by Bittiger et al. (1990).

cGMP determination

cGMP assays were performed using a radioimmunoassay kit with [^{3}H]cGMP obtained from Amersham (Amersham, Buckinghamshire, UK). Groups of 8 mice or rats were injected i.p. with test compounds or saline and sacrificed by fast focused microwave irradiation of the head (for mice: 3 sec, 2.8 kW, operating power; 2,450 MHz, 54 cm^{-2}; Medical Engineering Consultants, Lexington, MA) to prevent post mortem changes in levels of cGMP. Each cerebellum was dissected and homogenized by ultrasonication in 1 ml 0.05 M tris buffer with 4 mM EDTA, pH 7.5 (to prevent enzymatic degradation of cGMP), followed by heating 800 µl of the solution for 3 minutes at 120°C in a glycerine bath to coagulate protein. Homogenized samples were then centrifuged for 5 min at 40,000 × g in the cold. cGMP levels in 100 µl aliquots of the supernatants were assayed in duplicate with the radioimmunoassay kit. The procedure involved incubating [^{3}H]cGMP, antiserum and sample at 4°C for 1.5 to 18 hr. The antibody-cGMP complex was pelleted by the addition of chilled ammonium sulfate (60% saturated) and centrifugation. Pellets were resuspended in water, the suspension added to a scintillation cocktail, and radioactivity measured. Control experiments were carried out with an acetylated [^{125}I]cGMP RIA kit of Advanced Magnetic (Cambridge, MA).

Analysis of data

Results are expressed as means ± standard deviation for 6 to 10 animals per group. Dunnet's multiple comparison two-tailed test (Winer, 1971) was used to assess the significance of differences between several groups and Student's t-test for paired groups. Means ± SEM were considered to be statistically different when $p < 0.05$.

3. Results

Quantification, linearity, recovery and reproducibility

The GC characteristics and mass spectra of GBL are shown in Fig. 1A and 1B. The yield for the extraction of GBL using brain homogenates spiked with pure [2,3^{3}H]GHB was 100% (N = 8). Total recovery of the method, extraction plus derivatization, as estimated by adding different quantities of GHB (sodium salt) to brain extracts, was 100 ± 7.7%. The conversion of

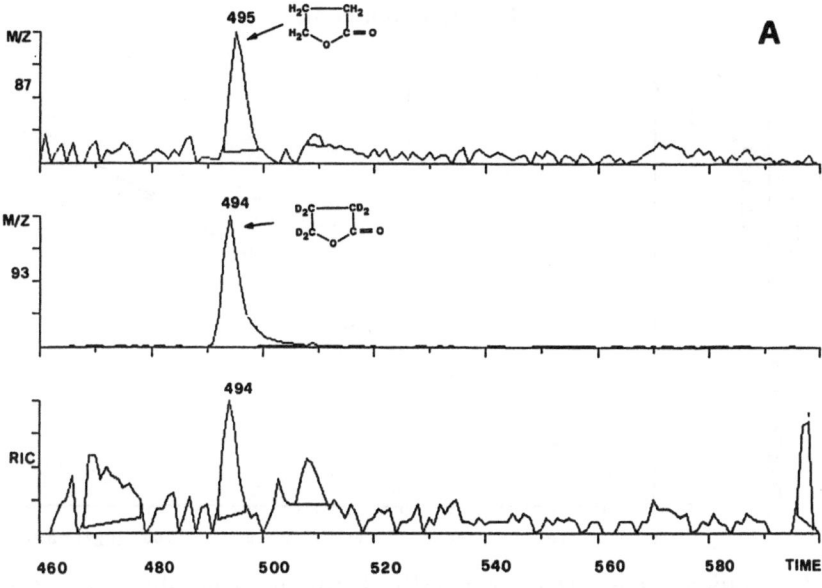

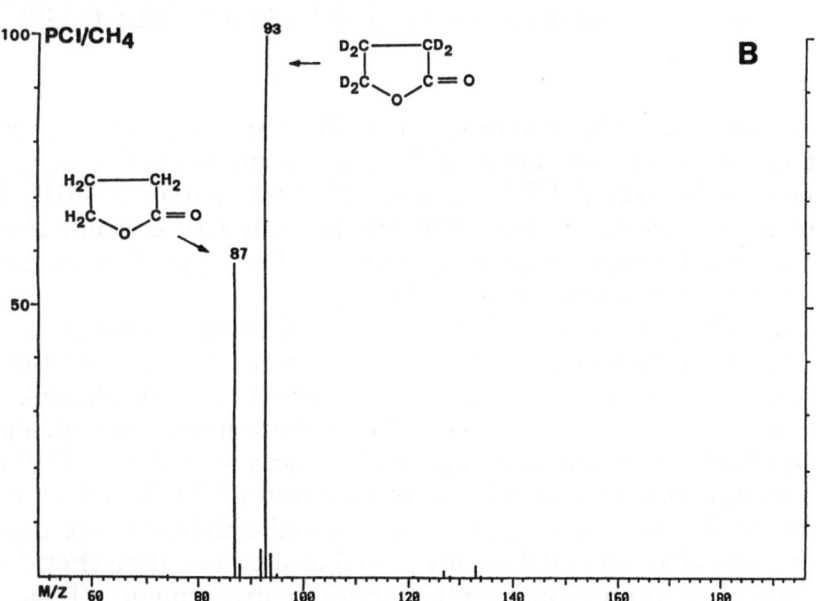

Fig. 1. Gas chromatography and mass spectra of GBL and GBL-d_6. These lactones were analyzed by extracting them from brain tissue with acetonitrile or with the combination acetonitrile and 5% acetic anhydride as described in "Materials and methods". 2 µl was injected into GC/MS system. GC conditions as described in "Materials and methods". **A** Mass spectrum of GBL-d_0 m/z = 87 and GBL-d_6 m/z = 93 used as internal standard for quantification. Ionization conditions are positive ion chemical ionization with methane as reagent gas. Time = retention time in sec. The number at the top of each peak represents the retention time of the corresponding lactone. *RIC* Reconstructed ion current. **B** GBL-d_0 m/z = 87 and GBL-d_6 m/z = 93 as internal standard selected ion mass chromatograms from brain extract. Ordinate = relative intensity in %

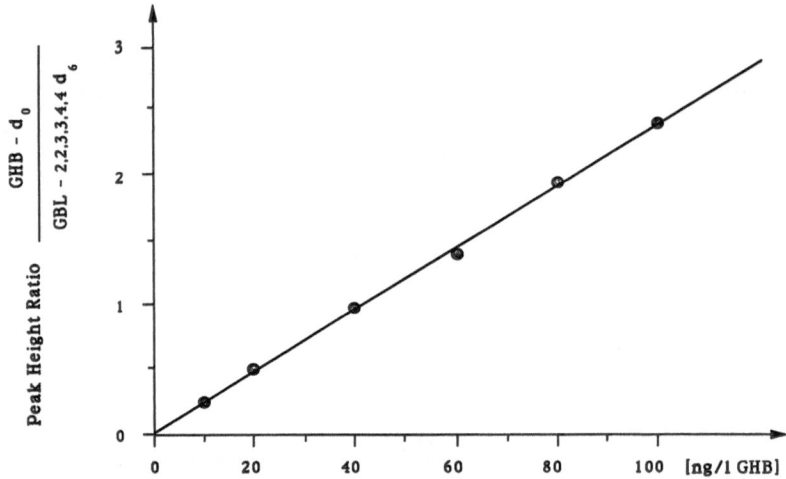

Fig. 2. Calibration curve GHB versus GBL-d$_6$ for the determination of GHB. Test samples containing various amounts of GHB and constant amounts of internal standard GBL-d$_6$ were derivatized to GBL-d$_0$ using conditions described in "Materials and methods" and injected into the gas chromatograph. Every point is the average of two determinations. Data are expressed as peak height ratio GHB-d$_0$: GBL-d$_6$

GHB to GBL under the conditions of acidification used was also quantitative. Standard curves were obtained by derivatizing quantities ranging from 5 to 100 ng GHB with 40 GBL-d$_6$. The calibration curves of GHB/internal standard peak area versus the GHB/internal standard concentration ratio showed a linear response in the range studied. The regression coefficient for the calibration curves was r $\geq$ 0.99 (Fig. 2).

The sensitivity of the assay is high, as the quantities of GHB-d$_6$ injected to get the fragmentograms of Fig. 1B are about 125 pg, and the setting of the electron multiplier is very low. In the total reproducibility assay (extraction, derivatization and GC-MS measurement), the quantities of GHB plus GBL measured from a pool of cortices were 2.36 $\pm$ 0.09 nmol/g, which corresponded to a coefficient of variation of 3.88% (N = 40). This variation coefficient is low because the internal standard is the deuterated derivative and GBL and GBL-d$_6$ are eluted at the same time (Fig. 1A). The mean cortical concentrations of GHB in all control samples (N = 57) was 2.12 $\pm$ 0.23 nmol/g, GBL was also found to be present in all brain structures investigated. The cortical concentration was 0.370 $\pm$ 0.025 nmol/g (N = 57). This is about 15% of the concentration of GHB in this brain area.

Extraction with organic solvents

A study of the most favorable conditions for the isolation and extraction of GBL and GHB from brain tissues was carried out with several organic

Table 1. Regional distribution of GHB and GBL in brain of rats

Brain areas	GBL levels nmol/g wet wt	GHB levels nmol/g wet wt
Cortex	0.37 ± 0.02 (57)	2.12 ± 0.23 (57)
Striatum	0.65 ± 0.06 (12)	4.67 ± 0.25** (12)
Hippocampus	0.39 ± 0.04 (38)	4.49 ± 0.91** (38)
Hypothalamus	N.D.	4.25 ± 0.48** (28)
Thalamus	N.D.	4.28 ± 0.90** (20)
Cerebellum	0.33 ± 0.02 (6)	2.33 ± 0.16 N.S. (6)

Each value represents the mean ± SEM for (N) rats. ND = not determined. Interregional significancies were estimated relative to cortical GHB level by the Student's t-test for paired groups. ** $p < 0.01$

solvents by adding GBL and GHB before the homogenisation of the irradiated tissue and analysis with the GC-MS method. Best recovery (>95%) was obtained with acetonitrile. The acetonitrile extracts yield much cleaner chromatograms than do extracts prepared from other organic solvents (ethanol, methanol, chloroform, dioxane and tetrahydrofurane).

Assay of GHB and GBL in different brain structures

The amounts of GHB and GBL in 6 regions of the rat brain are indicated in Table 1. The structures richest in GHB are striatum, hypothalamus, hippocampus and thalamus, whereas cortex and cerebellum had a relatively

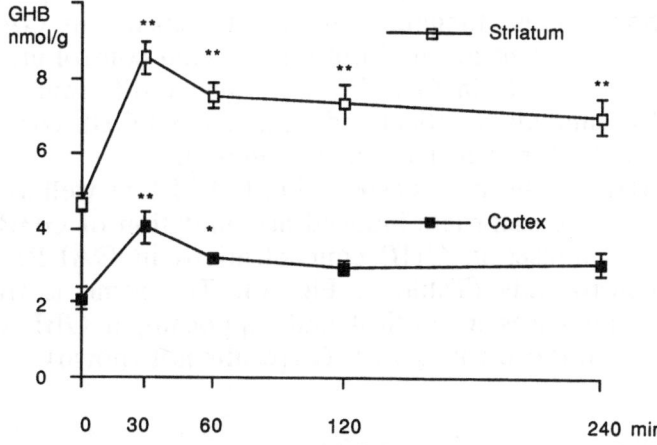

Fig. 3. Time course of GHB accumulation in cortex and striatum of rats treated with valproate (400 mg/kg i.p.). Results are the mean ± S.E.M. for six animals. Statistical significance was calculated by Dunnett's test: * $p < 0.05$, ** $p < 0.01$ when compared to the control group at t = 0. The initial rate of GHB synthesis in the cortex and striatum were 3.84 nmol/g/h and 7.78 nmol/g/h, respectively

low content of GHB. Statistically significant differences ($p < 0.01$; Student's t test for paired group) were observed between the cortex (= 100%) versus the following areas: striatum, 220%; hippocampus 210%; hypothalamus 200% and thalamus 200%. In general, the distribution pattern for GBL seems to follow that of GHB.

Time course of GHB accumulation in rats treated with valproate

Rats were treated with valproate (400 mg/kg i.p.), killed by microwave irradiation 0, 30, 60, 120 and 240 minutes later; GBL and GHB levels were determined in cortex and striatum. In these two regions, valproate induced a rapid and strong increase of GHB and GBL levels (about 180%) for 30 minutes; then the content of the two GABA metabolites decreased slightly until a plateau was reached (Fig. 3). This rapid accumulation of GHB observed 30 minutes after enzymatic inhibition of its metabolization was used to determine the rate of GHB synthesis by calculating the difference between GHB content 30 minutes after treatment with valproate and the control level. In the different regions investigated, the accumulation of GBL caused by valproate was of the same order of magnitude as for GHB (180%); but these increases never reached the level of significance (results not shown).

GHB and GBL content and rate of synthesis in rats with SWD compared to controls

The concentrations of endogenous GHB and of its prodrug GBL were measured in hippocampus, thalamus and frontal cortex in GAERS and compared to those observed in rats from the selected control group (Table 2, Fig. 4). Levels of GHB in GAERS were never different from those observed in control animals. Cortical and hippocampal GBL concentrations were also similar in both strains (results not shown).

The rate of GHB synthesis was assessed in GAERS as well as in control rats by reference to the valproate-induced accumulation of GHB (Table 2 and Fig. 4). The increases in GHB concentrations in GAERS were not different from control rats (Table 2, Fig. 4). The same is true of the valproate-induced increases in cortical and hippocampal GBL content in rats with SWD and in those without SWD (results not shown).

Selectivity of interactions of GHB with $GABA_B$ receptors

GHB interacted with the $GABA_B$ receptors with an IC_{50} of 1.5×10^{-4}M. This value was obtained in three different experiments using [³H]baclofen as radioligand and membranes prepared from cerebral cortices according to

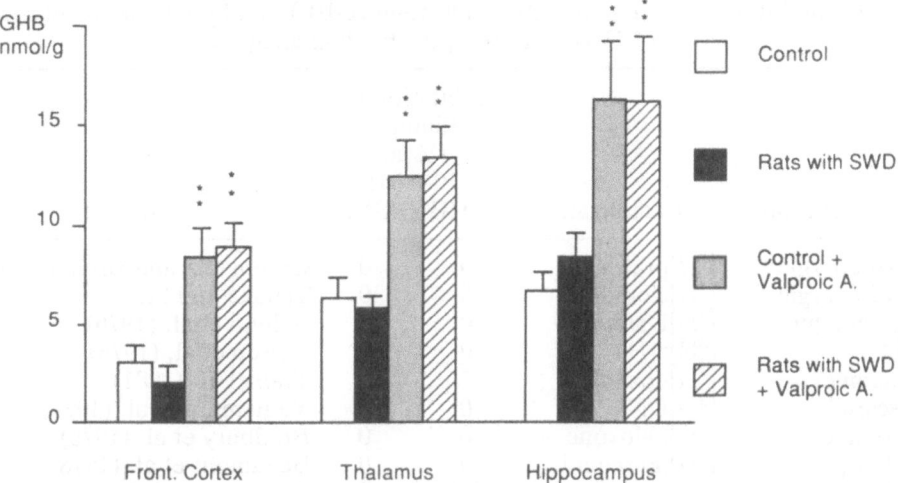

Fig. 4. GHB levels and GHB rate of synthesis in rats with SWD as compared to "non-epileptic" control rats. Animals treated with valproate were sacrificed 30 min later. Results are the mean ± S.E.M. for groups of ten rats. **p < 0.01 when compared to the respective control group (Dunnett's test)

Table 2. Kinetic parameters for the synthesis of GHB in rats with SWD as compared to controls

Brain areas	Control GHB content nmol/g	GHB content after valproate nmol/g	Initial rate of GHB synthesis nmol/g/h	Turnover time h
Cortex GAERS	2.02 ± 0.83	8.90 ± 1.14**	13.76	0.15
Cortex control	3.06 ± 0.84	8.31 ± 1.54**	10.50	0.29
Hippoc.GAERS	8.34 ± 1.15	16.17 ± 3.34**	15.68	0.53
Hippoc.control	6.67 ± 0.91	16.30 ± 2.86**	19.24	0.35
Thalam.GAERS	5.69 ± 0.71	13.38 ± 1.46**	15.38	0.37
Thalam.control	6.25 ± 1.05	12.47 ± 1.83**	12.44	0.50

Wistar rats from the colony of Strasbourg (GAERS) were treated with valproate, killed 30 min later and GHB levels were determined in dissected brain regions. Control GHB concentrations were determined in animals receiving saline. All values are means ± S.E.M. for 10 animals per group and refer to wet weight.
Statistical significance of difference was calculated by Dunnet's test: **p < 0.01

Bernasconi et al. (1986). Interactions with other receptors (including $GABA_A$ and central benzodiazepine receptors) were absent at a concentration of 100 μM (Table 3). GBL did not interact with the 12 receptors listed in Table 3, including $GABA_B$ receptors, at a concentration of 100 μM. Thus, the interaction of GHB with $GABA_B$ receptors appears to be selective.

For the measurement of the interaction of GHB with $GABA_B$ receptors in GAERS as compared to control rats, the potent and highly selective

R. Bernasconi et al.

Table 3. Inhibition of binding by γ-butyrolactone (GBL) and by γ-hydroxybutyric acid (GHB) in 12 receptor binding assays

Putative receptor	Radioligand	Inhibition of binding (% at 10^{-4} M) GHB	GBL	Method
α_1-Adrenergic	[^{3}H]prazosin	0	0	Greengrass and Bremner (1979)
α_2-Adrenergic	[^{3}H]clonidine	0	0	Tanaka and Starke (1980)
β-Adrenergic	[^{3}H]DHA	0	0	Bylund et al. (1976)
5-HT$_1$	[^{3}H]5-HT	0	0	Nelson et al. (1978)
Histamine$_1$	[^{3}H]doxepine	5*	0	Tran et al. (1981)
Muscarinic	[^{3}H]QNB	0	0	Yamamura et al. (1974)
Mu-opiate	[^{3}H]naloxone	0	0	Bradbury et al. (1976)
GABA$_A$	[^{3}H]muscimol	0	0	Beaumont et al. (1978)
GABA$_B$	[^{3}H]baclofen	44**	0	Bernasconi et al. (1986)
Benzodiazepine	[^{3}H]flunitrazepam	0	0	Speth et al. (1978)
Adenosine A1	[^{3}H]CHA	0	0	Patel et al. (1982)
Substance P	[^{3}H]substance P	0	0	Bittiger et al. (1982)

The receptor binding assays were performed essentially as described in the references. Abbreviations: *DHA* dihydro-alprenolol; *5-HT* serotonin; *QNB* quinuclidinyl benzylate; *CHA* cyclohexyl-adenosine. * =55% at 10^{-3} M; ** IC$_{50}$ = 1.5 × 10^{-4} M obtained from 3 inhibition curves

Table 4. Interactions of γ-hydroxybutyric acid with GABA$_B$ receptors in rats with SWD and in rats from the selected control group
IC$_{50}$ in μM

Brain structures	Control rats	Rats with SWD
Cortex	152.5	132.2
Cerebellum	138.0	168.4
Thalamus	166.7	157.2

Membranes were prepared from male Wistar rats from the breeding colony at the Centre de Neurochimie, C.N.R.S., Strasbourg according to Bittiger et al. (1990). The radioreceptor assay was performed with [^{3}H]CGP 27492 as radioligand according to Bittiger et al. (1990)

GABA$_B$ radioligand, [^{3}H]CGP 27492, was used (Bittiger et al., 1988, 1990). The IC$_{50}$ ranged from 1.38 × 10^{-4} M in the cerebellum to 1.66 × 10^{-4} M in the thalamus and were similar in GAERS and in control rats (Table 4) and not different from the IC$_{50}$ values obtained with [^{3}H]baclofen as radioligand and membranes prepared from cerebral cortices (Table 4).

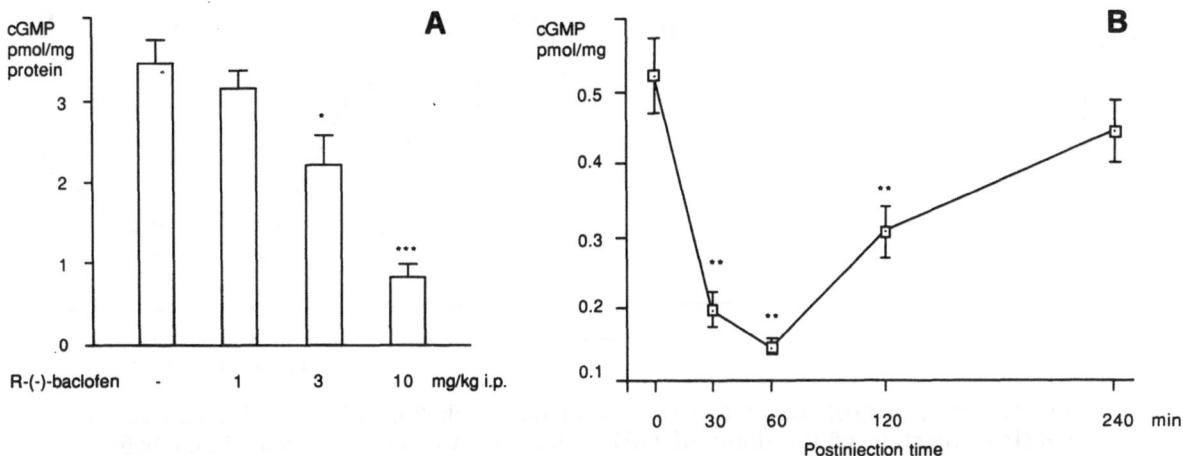

Fig. 5. Cerebellar cGMP concentrations in mice exposed to R(−)-baclofen. **A** Dose-dependent decrease in cGMP content. Animals (n = 8) were administered R(−)-baclofen and killed 60 min later, controls received 0.9% saline. cGMP levels were determined by radioimmunoassay and expressed as mean ± S.E.M. **p < 0.01, ***p < 0.001 (Dunnett's test). **B** Time course of cGMP levels in the cerebellum following administration of R(−)-baclofen (6 mg/kg, i.p.). Each value represents the mean ± S.E.M. of 8 mice. Controls received 0.9% saline and were killed 30 min later

Effects of GHB and the GABA_B receptor agonist R(−)-baclofen on cerebellar cGMP content

The GABA_B receptor agonist R(−)-baclofen dose-dependently decreased cerebellar cGMP levels (Fig. 5A). The threshold dose 60 min after injection was between 1 and 3 mg/kg i.p. (56% of control at 3 mg/kg) and the content of cGMP after 6 mg/kg R(−)-baclofen was 28% of control value and decreased to 20% at 10 mg/kg. Figure 5B shows the time-course for the decrease of cerebellar cGMP observed after injection of 6 mg/kg of the agonist. The onset of the decrease of cGMP content caused by R(−)-baclofen was very rapid; 30 min after administration of R(−)-balcofen cGMP levels were 38% of controls and decreased further to 28% 60 min after drug treatment. Then, levels of the second messenger increased again and reached 59% of control values 2 hours after administration of R(−)-baclofen. After 4 hours cerebellar cGMP concentrations were normalized and ataxia had disappeared in mice. This suggests that the behavioural effects induced by R(−)-baclofen correlate with the decrease in cGMP.

GBL decreased cGMP levels in a dose-dependent manner (Fig. 6A). While 100 mg/kg GBL i.p. did not alter cerebellar cGMP levels significantly 45 min after administration of the drug (74%), the reductions by 200 mg/kg GBL i.p. (42%) and 400 mg/kg GBL i.p. (24%) were statistically significant (p < 0.01). The time course of the effect of GBL on levels of cerebellar cGMP is shown in Fig. 6B. After intraperitoneal administration of

168 R. Bernasconi et al.

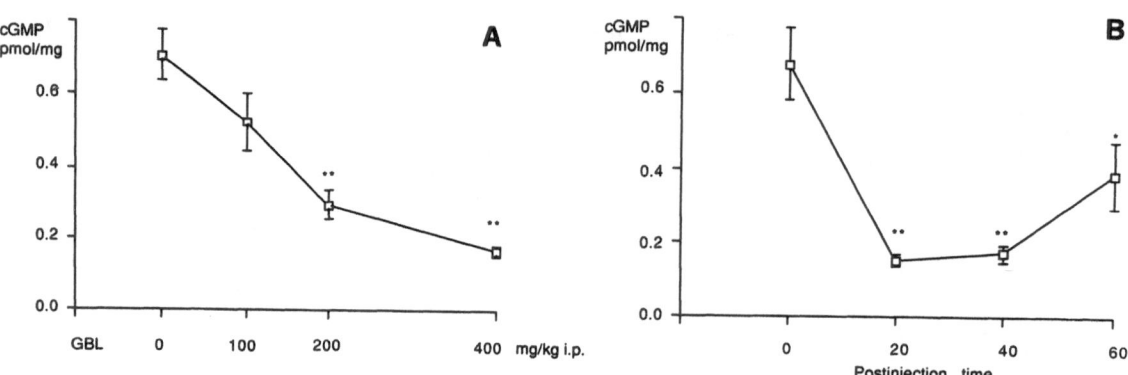

Fig. 6. Effect of GBL on cGMP content in the cerebellum of mice. **A** Decrease in cGMP in function of the doses of GBL. The drug was administered 45 min before microwave irradiation. Each value represents the mean ± S.E.M. of 8 animals. Statistical significance was calculated by Dunnett's test. **B** Time course of cGMP concentrations in the cerebellum of mice following injection of 200 mg/kg GBL. Each point represents the mean ± S.E.M. of 8 animals. $^*p < 0.05$, $^{**}p < 0.01$

200 mg/kg of GBL, there was a rapid and marked decrease in the content of cGMP, the maximal effect being achieved after 20 min (22%). This effect lasted up to 40 min (25%); then levels of cGMP increased slightly and reached 57% of control value after 60 min. Similar results were also observed in cerebellum and thalamus of rats after treatment with 3.5 mmol/kg GBL (results not shown).

Antagonism of the decrease of cGMP caused by R(−)-baclofen and GBL by the GABA$_B$ receptor antagonist CGP 35348

The GABA$_B$ receptor antagonist CGP 35348 at doses of 100 and 200 mg/kg i.p. did not alter cerebellar cGMP levels of mice; at 400 mg/kg a slight (about 20%) and occasionally significant increase of cGMP was observed (results not shown). However, the 60% decrease of cGMP content induced by 4 mg/kg i.p. R(−)-baclofen was completely antagonized by 200 mg/kg i.p. CGP 35348 (Fig. 7).

The marked decrease (35% of control value) of cerebellar cGMP concentration observed in mice 40 min after injection of 200 mg/kg i.p. GBL was significant, but only partially antagonized (58% of control value, = 36% antagonism, N = 3) by 200 mg/kg i.p. CGP 35348 given 5 min before GBL (Fig. 8). Similar results were observed in mice when GBL (200 mg/kg) was injected 40 min after 200 mg/kg CGP 35348 and the animals were killed 20 min later by microwave irradiation (results not shown). In addition, the decrease of cGMP caused by 200 mg/kg GBL was also only partially, but significantly, antagonized by pretreatment with the high dose of 400 mg/kg CGP 35348 (CGP 35348 was given 30 min before GBL and 50 min before

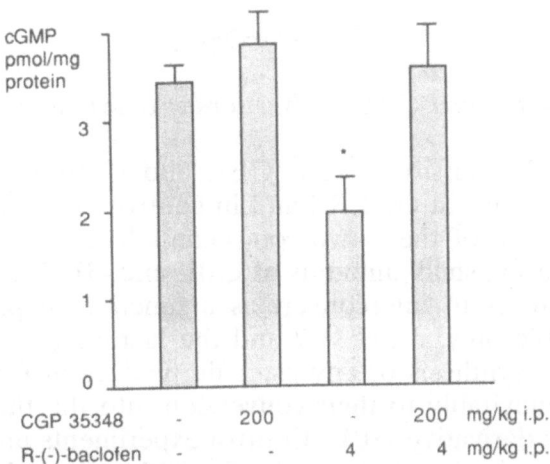

Fig. 7. Antagonism by CGP 35348 of the decrease in levels of cGMP in the cerebellum induced by R(−)-baclofen. CGP 35348 was injected 20 min before R(−)-baclofen and mice were sacrificed 60 min after treatment with the $GABA_B$ receptor antagonist. Each group represents the mean ± S.E.M. of 8 animals. *p < 0.01

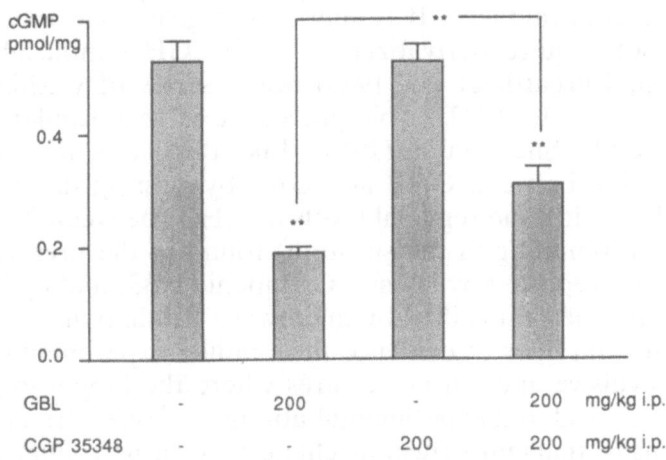

Fig. 8. Partial antagonistic effect of CGP 35348 on the decrease of cerebellar cGMP produced by GBL. CGP 35348 was injected 5 min before GBL and mice were killed 40 min later. The results are the mean ± S.E.M. of duplicate determination. **p < 0.01 (Dunnett's test)

sacrifice, control values = 100%, CGP 35348 = 123% N.S., GBL = 23%*, CGP 35348 + GBL = 35%*,#; *p < 0.01 versus control, #p < 0.05 versus GBL group). A partial antagonism of the decrease of cerebellar cGMP was observed in rats when 3.5 mmol/kg GBL were injected 30 min after 200 mg/kg i.p. CGP 35348 and animals sacrificed 60 min after the administration of the $GABA_B$ receptor antagonist (results not shown).

4. Discussion

GHB and GBL levels and rates of synthesis

In order to test the possibility that GBL and GHB may be present at different concentrations in GAERS and in control rats, a sensitive method for the determination of these two compounds had to be developed. The difficulty of measuring small amounts of GBL and GHB lies in the capacity of the two substances to interconvert as a function of pH, the free acid (GHB) being stable only at pH > 7 and the lactone (GBL) and pH < 7. Consequently, the synthesis of any ester derivatives of GHB under acidic conditions leads inevitably to their conversion into the thermodynamically more stable cyclic derivative GBL. Control experiments under various conditions of the derivatization, followed by GC-MS analysis demonstrated that the yields of ester derivatives of GHB were less than 1% (J. Lauber, unpublished results). We utilized this characteristic of the GHB/GBL system and measured only GBL, that was present normally, and that which was formed from GHB.

The use of the GC/MS assay procedure for GHB and GBL described herein has several advantages. It is simple and gives high recovery yields since it does not require derivatization of the GHB molecule (Eli and Cattabeni, 1983; Ehrhardt et al., 1988) nor a series of washings and re-extraction (Snead et al., 1982). This procedure is very similar to the one recently described by Snead et al. (1989). Like this latter method, it allows clear differentiation between GHB and GBL by altering the conditions of extraction of the brain. The regional levels of GHB (between 2.12 nmol/g in the cortex and 4.67 nmol/g in the striatum) found in the present study are close to the values reported by Eli and Cattabeni (1983) and by Vayer et al. (1988) for age-matched rats killed by microwave irradiation.

There are several lines of evidence that, unlike experimental models of generalized convulsive and partial seizures where the hippocampus plays a dominant role, clinical and experimental absence seizures are characterized by perturbations of thalamocortical mechanisms with no significant involvement of the hippocampus (Avoli and Gloor, 1982; Gloor and Fariello, 1988; Steriade and Llinas, 1988; Vergnes et al., 1987, 1990). If an alteration of GHB and GBL levels, or of their rate of synthesis, is implicated in the pathogenesis of petit-mal epilepsy it should in all probability occur in the frontal cortex and in the thalamus, and no effect should be observed in the hippocampus.

Tissue concentration of GHB in the cortex at the onset of SWD induced with threshold doses of either 3.5 mmol/kg i.p. GHB or GBL was reported by Snead (1991) to be 240 ± 31 nmol/g. This threshold cortical concentration is one hundredfold higher than the naturally occurring content found in this and earlier studies. Thalamic GHB levels were not reported in Snead's study. Assuming that GHB is disributed more or less evenly throughout the frontal cortex and thalamic nuclei, and that high local concentrations do not exist it would seem unlikely that the endogenous concentrations observed in the present study could exert a significant inhibitory effect capable of inducing SWD. This study has also demonstrated that GHB and GBL levels measured in the three structures were not different in GAERS from the levels measured in control rats (Fig. 4). These findings suggest that endogenous GHB or GBL concentrations do not play an important role in the pathogenesis of absence epilepsy. This hypothesis is also supported by the fact that ethosuximide and valproate have opposite effects on endogenous GHB levels of rat brains. The concentration of GHB decreases after chronic treatment with ethosuximide (Snead et al., 1980), but dose-dependently increases after administration of valproate (Vayer et al., 1988; this study).

Measurement of the rate of GHB synthesis would be a more reliable index of GHBergic functional activity than assessment of regional concentrations. The first observation of an accumulation of GHB following acute treatment with valproate was reported by Snead et al. (1980), and Vayer et al. (1988) used this model to estimate GHB turnover rates in discrete brain regions of the rat. Our results confirm their findings and describe regional variations in the kinetics of these accumulations. The initial rates of GHB and GBL synthesis observed in GAERS and rats from the selected control group are of the same order of magnitude as those reported by Vayer et al. (1988). The data of this study are also in good accordance with the turnover time of GHB in total brain (26 min) determined by Gold and Roth (1977) using [^{3}H]GABA as the GHB precursor.

This increase of GHB levels is due to the fact that valproate does not alter the biosynthesis of GHB, but inhibits its degradation, causing a rapid accumulation of the substance (Vayer et al., 1988). GHB accumulation does not induce a feedback inhibition of GHB synthesis, since no significant effect of GHB has been observed on its synthetic enzyme (Rumigny et al., 1980). Thus, the use of valproate for the determination of GHB turnover seems to be justified. However, these measurements do not necessarily reflect the turnover rate of GHB in the synapses, but rather the overal rate of GHB synthesis in brain.

The rate of GHB synthesis in rats in GAERS was not significantly different from the one observed in rats from the selected control group. These data indicate that the pharmacological basis for the mechanism underlying the occurrence of cortical SWD in the genetic model of absence seizures developed in Strasbourg lies elsewhere, and Snead et al. (1990) have reported an increase in the density of low affinity GHB binding sites in GAERS. However, this observation does not exclude other mechanisms.

Interactions between GHB and GABA$_B$ receptors

As CGP 35348 antagonized the SWD caused by GHB (Marescaux et al., 1992) we tested the affinity of GHB and GBL for GABA$_A$, GABA$_B$ and benzodiazepine receptors. We observed a selective, but weak affinity of GHB for $GABA_B$ receptors. Similar IC_{50} values of GHB for displacing [^{3}H]CGP 27492 from $GABA_B$ binding sites were observed in three different brain structures in GAERS and in control rats (Table 4). Analysis of the binding data is consistent with a single binding site and does not suggest allosteric interactions between GHB and $GABA_B$ receptors.

An IC_{50} value of the order of 140 µm for GHB and brain levels of 2.12 nmol/g GHB suggest that GHB stimulation of $GABA_B$ receptors is not the mechanism underlying the pathogenesis of SWD in GAERS. A functional involvement of GHB in the pathogenesis of GAERS would appear improbably unless GHB were highly concentrated locally (about 200 nmol/g) in brain structures such as the lateral thalamus from where SWD emanate, and in view of the average thalamic concentration of 4 nmol/g found in this and other studies that seems unlikely.

On the other hand, according to Snead (1991), the cortical GHB concentrations at the onset of SWD after injection of the threshold dose of 3.5 mmol/kg i.p. GBL are 240 ± 31 nmol/g, i.e., sufficient to stimulate $GABA_B$ receptors. Thus, non-anaesthetic doses of GBL could stimulate $GABA_B$ receptors, and the generation of SWD in the GHB model of absence epilepsy might be due to the $GABA_B$ receptor agonist-like properties of GHB.

These results suggest that $GABA_B$ receptors could assume an important role in absence epilepsy. A dose-related increase in SWD after stimulation of $GABA_B$ receptors with R(−)-baclofen was first described by Vergnes et al. (1984). In human studies, absence attacks have been found to occur more frequently during treatment with baclofen (Gloor and Fariello, 1988).

As $GABA_A$ receptor agonists such as muscimol and THIP exacerbate experimental absence seizures, it was originally thought that $GABA_A$-mediated mechanisms must be involved in the pathogenesis of absences. $GABA_A$ receptor antagonists do not block experimental absence (Micheletti et al., 1985). Because of these negative results and because of the aggravating effects of R(−)-baclofen in experimental absences, we tested the novel, centrally active $GABA_B$ receptor antagonist, CGP 35348, in several animal model of absence seizures. CGP 35348 markedly and dose-dependently decreases SWD in GAERS and antagonizes the aggravating effects of agents which enhance GABA-mediated inhibition (vigabatrin, muscimol, THIP, SKF 89976). In addition, SWD induced by GHB were also antagonized by this $GABA_B$ receptor antagonist (Marescaux et al., 1992; Snead, personal communication). These results demonstrate that $GABA_B$ receptors are primarily involved in the genesis of SWD (Marescaux et al., 1992) and confirm the hypothesis of Crunelli and Leresche (1991), that activation of $GABA_B$ receptors mediates a late and long-lasting inhibitory postsynaptic potential (IPSP) which is critical for the generation of SWD.

The evaluation of the functional relevance of the in vitro binding results requires in vivo paradigms in which the effects of GHB on $GABA_B$ receptors can be demonstrated. The alterations of cerebellar cGMP levels induced by agonists and antagonists of several receptors, and in particular of $GABA_B$ receptors, have been used as a biochemical index of the signal transduction (Wood, 1991). Recent autoradiographic studies have demonstrated a clear topographic $GABA_B$ receptor distribution with parasagittal zones of high and low density of binding (Albin and Gilman, 1989; Bowery et al., 1987). These cerebellar $GABA_B$ receptors, when activated, decrease cGMP levels (Gumulka et al., 1979). We have confirmed the findings of Gumulka et al. demonstrating that the potent and selective $GABA_B$ receptor agonist R(−)-baclofen decreases cerebellar cGMP in a dose- and time-dependent fashion (Fig. 5A and 5B). GBL produces the same effect (Fig. 6A and 6B).

In this biochemical paradigm, R(−)-baclofen and GHB are roughly equal in efficacy, although the decrease observed after injection of GHB occurs faster than that induced by R(−)-baclofen. The ratio of the doses of R(−)-baclofen (6 mg/kg i.p.) and GBL (400 mg/kg i.p.) needed to bring about the same decrease in cGMP does not correspond to ratio of IC_{50} binding values (30 nM for baclofen as against 140 µM for GHB). This discrepancy may be due to better penetration of GBL into the brain and/ or to the fact that GHB interacts not only with $GABA_B$ receptors, but also with several neurotransmitter systems, including the noradrenergic, dopaminergic, serotonergic and cholinergic systems (Snead, 1977). All these neurotransmitter systems have been implicated in the control of cerebellar cGMP levels (Wood, 1991). Hence, GHB could exert an effect through partly different neurochemical systems to produce SWD and to decrease cerebellar cGMP content.

This led us to check whether the decrease in cGMP levels induced by GBL and R(−)-baclofen could be solely attributable to $GABA_B$ receptor activation. CGP 35348 should differentiate between effects of these two compounds that are related and those not related to the $GABA_B$ receptor. The results shown in Fig. 7 demonstrate that 200 mg/kg i.p. CGP 35348 fully antagonizes the decrease of cerebellar cGMP produced by 4 mg/kg i.p. R(−)-baclofen. The hitherto available evidence indicated that the dose of CGP 35348 needed to suppress the SWD induced by administration of 4 mg/kg i.p. R(−)-baclofen was also 200 mg/kg i.p. (Marescaux et al., 1992). This relation between the doses of the $GABA_B$ receptor agonist and antagonist is the same in the two experiments. By contrast, CGP 35348, which by itself has no effect on cGMP levels at doses up to 200 mg/kg i.p., only partially antagonizes the decrease caused by 200 mg/kg i.p. GBL at 200 mg/kg (Fig. 8) and at the very high dose of 400 mg/kg i.p. (results not shown). Thus, CGP 35348 blocked all the effects caused by the $GABA_B$ receptor agonist R(−)-baclofen, but totally antagonized only the increase in the duration of SWD induced by GHB, indicating that these effects may be $GABA_B$ receptor mediated. The decrease in cerebellar cGMP concentrations elicited by GHB was only partially antagonized by CGP 35348 and is therefore only partially mediated through $GABA_B$ receptors.

In sum, the results of this study do not support the hypothesis that endogenous GHB plays an important role in the pathogenesis and control of generalized absence seizures in GAERS. The cortical levels of GHB reached after administration of GBL are likely to stimulate $GABA_B$ receptors and thus to induce SWD. These findings, in conjunction with those of Marescaux et al. (1992) underline the primary importance of $GABA_B$ receptors in experimental absence epilepsy and in the genesis and control of spontaneous SWD in GAERS (Crunelli and Leresche, 1991). $GABA_B$ receptor antagonists may represent a new class of anti-absence drugs.

Acknowledgements

We are indebted to the many people who offered critical comments and helpful suggestions upon reading earlier versions of this manuscript, including W. Froestl, K. Hauser, L. Maitre, H. R. Olpe and P. Waldmeier.

References

Albin RL, Gilman S (1989) Parasagittal zonation of $GABA_B$ receptors in molecular layer of rat cerebellum. Eur J Pharmacol 173: 113–114

Avoli M, Gloor P (1982) Interaction of cortex and thalamus in spike and wave discharges of feline generalized penicillin epilepsy. Exp Neurol 76: 196–217

Beaumont K, Chilton WS, Yamamura HI, Enna SJ (1978) Muscimol binding in rat brain: association with synaptic GABA receptors. Brain Res 148: 153–162

Bernasconi R, Jones RSG, Bittiger H, Olpe HR, Heid J, Martin P, Klein M, Loo P, Braunwalder A, Schmutz M (1986) Does pipecolic acid interact with the central GABA-ergic system? J Neural Transm 67: 175–189

Bittiger H (1982) Substance P receptors in the nervous system and possible receptors subtypes (discussion). In: Porter R, O'Connor M (eds) Substance P in the nervous system. Ciba Foundation symposium 91. Pitman, London, pp 196–201

Bittiger H, Reymann N, Hall R, Kane P (1988) CGP 27492, a new, potent and selective radioligand for $GABA_B$ receptors. In: Proceedings of the 11th Annual Meeting of the European Neuroscience Association. Zürich, September 1988. Eur J Neurosci [Suppl]: Abstr. 16.10

Bittiger H, Froestl W, Hall R, Karlsson G, Klebs K, Olpe HR, Pozza M, Steinmann M, Van Riezen H (1990) Biochemistry, electrophysiology and pharmacology of a new $GABA_B$ antagonist: CGP 35348. In: Bowery N, Bittiger H, Olpe HR (eds) $GABA_B$ receptors in mammalian function. J Wiley, Chichester, pp 47–60

Bowery NG, Hudson AL, Price GW (1987) $GABA_A$ and $GABA_B$ receptor site distribution in the rat central nervous system. Neuroscience 20: 365–383

Bradbury AF, Smyth DG, Snell CR, Birdsall NJ, Hulme EC (1976) C-fragment of lipotropin has a high affinity for brain opiate receptors. Nature 260: 793–795

Bylund DB, Snyder SH (1976) Beta adrenergic receptor binding in membrane preparation from mammalian brain. Mol Pharmacol 12: 568–580

Crunelli V, Leresche N (1991) A role for $GABA_B$ receptors in excitation and inhibition of thalamocortical cells. Trends Neurosci 14: 16–21

Eli M, Cattabeni F (1983) Endogenous γ-hydroxybutyrate in rat brain areas: post-mortem changes and effects of drugs interfering with γ-aminobutyric acid metabolism. J Neurochem 41: 524–530

Ehrhardt JD, Vayer P, Maitre M (1988) A rapid and sensitive method for the determination of γ-hydroxybutyric acid and trans-γ-hydroxycrotonic acid in rat brain tissue by gas chromatography/mass spectrometry with negative ion detection. Biomed Environ Mass Spectrom 15: 521–524

Engel J, Ochs RF, Gloor P (1990) Metabolic studies of generalized epilepsy. In: Avoli M, Gloor P, Kostopoulos G, Naquet R (eds) Generalized epilepsy: neurobiological approaches. Birkhäuser, Boston, pp 387–396

Enna SJ, Snyder SH (1975) Properties of gamma-aminobutyric acid (GABA) receptor binding in ra brain synaptic membrane fractions. Brain Res 100: 81–97

Fariello RG, Golden GT (1987) The THIP-induced model of bilateral synchronous spike and wave in rodents. Neuropharmacology 26: 161–165

Fromm GH, Kohli CM (1972) The role of inhibitory pathways in petit mal epilepsy. Neurology 22: 1012–1020

Gloor P, Fariello RG (1988) Generalized epilepsy: some of its cellular mechanisms differ from those of focal epilepsy. Trends Neurosci 11: 63–68

Glowinski J, Iversen LL (1966) Regional studies of catecholamines in the rat brain-I. The disposition of [^{3}H]norepinephrine, [^{3}H]dopamine and [^{3}H]DOPA in various regions of the brain. J Neurochem 13: 655–669

Godschalk M, Dzoljic MR, Bonta IL (1976) Antagonism of gamma-hydroxybutyrate-induced hypersynchronization in the ECoG of the rat by anti-petit mal drugs. Neurosci Lett 3: 145–150

Godschalk M, Dzoljic MR, Bonta IL (1977) Slow wave sleep and a state resembling absence epilepsy induced in the rat by γ-hydroxybutyrate. Eur J Pharmacol 44: 105–111

Gold BL, Roth RH (1977) Kinetics of in vivo conversion of gamma [^{3}H]aminobutyric acid to gamma [^{3}H]hydroxybutyric acid in rat brain. J Neurochem 28: 1069–1073

Greengrass P, Bremner R (1979) Binding characteristics of ^{3}H-prazosin to rat brain alpha-adrenergic receptors. Eur J Pharmacol 55: 323–326

Grob K (1986) Making and manipulating capillary columns for gas chromatography. Huethig, Basel New York

Gumulka SW, Dinnendahl V, Schonhoffer PS (1979) Baclofen and cerebellar cyclic GMP levels in mice. Pharmacology 19: 75–81

King GA (1979) Effects of systemically applied GABA agonists and antagonists on wave-spike ECoG activity in rat. Neuropharmacology 18: 47–55

Marescaux C, Micheletti G, Vergnes M, Depaulis A, Rumbach L, Warter JM (1984) A model of chronic spontaneous petit mal-like seizures in the rat: comparison with pentylenetetrazol-induced seizures. Epilepsia 25: 326–331

Marescaux C, Vergnes M, Bernasconi R (1992) GABA$_B$ receptor antagonists: potential new anti-absence drugs (this volume)

Meldrum B (1981) GABA-agonists as anti-epileptic agents. In: Costa E, Di Chiara G, Gessa GL (eds) GABA and benzodiazepine receptors. Adv Biochem Psychopharmacol 26: 207–217

Micheletti G, Marescaux C, Vergnes M, Rumbach L, Warter JM (1985) Effects of GABA-mimetics and GABA antagonists on spontaneous nonconvulsive seizures in Wistar rats. In: Bartholoni G, Bossi L, Lloyd KG, Morselli PL (eds) Epilepsy and GABA receptor agonists. Raven Press, New York, pp 129–137

Mirsky AF, Duncan CC, Myslobodsky MS (1986) Petit mal epilepsy: a review and integration of recent information. J Clin Neurophysiol 3: 179–208

Nelson PL, Herbert A, Bourgoin S, Glowinski J, Hamon M (1978) Characteristics of central 5-HT receptors and their adaptive changes following intracerebral 5,7-dihydroxytryptamine administration in the rat. Mol Pharmacol 14: 983–995

Patel J, Marangos PJ, Stivers J, Goodwin FK (1982) Characterization of adenosine receptors in brain using N6 cyclohexyl [^{3}H]adenosine. Brain Res 237:203–214

Pericic D, Eng N, Walters JR (1978) Post-mortem and aminooxyacetic acid-induced accumulation of GABA: effect of gamma-butyrolactone and picrotoxin. J Neurochem 30: 767–773

Rumigny JF, Maitre M, Cash C, Mandel P (1980) Specific and non-specific succinic semialdehyde reductases from rat brain: isolation and properties. FEBS Lett 117: 111–116

Smith KA, Bierkamper GG (1990) Paradoxical role of GABA in a chronic model of petit mal (absence)-like epilepsy in the rat. Eur J Pharmacol 176: 45–55

Snead OC (1977) Gamma hydroxybutyrate. Life Sci 20: 1935–1943

Snead OC (1988) γ-Hydroxybutyrate model of generalized absence seizures: further characterization and comparison with comparison with other absence models. Epilepsia 29: 361–368

Snead OC (1990) The ontogeny of GABAergic enhancement of the γ-hydroxybutyrate model of generalized absence seizures. Epilepsia 31: 363–368

Snead OC (1991) The γ-hydroxybutyrate model of absence seizures: correlation of regional brain levels of γ-hydroxybutyric acid and γ-butyrolactone with spike wave discharges. Neuropharmacology 30: 161–167

Snead OC, Yu RR, Huttenlocher RR (1976) γ-Hydroxybutyrate: correlation of serum and cerebrospinal fluid levels with electroencephalographic and behavioural effects. Neurology 26: 51–56

Snead OC, Bearden LH, Pegram V (1980) Effect of acute and chronic anticonvulsant administration on endogenous gamma-hydroxybutyrate in rat brain. Neuropharmacology 19: 47–52

Snead OC, Liu CC, Bearden LJ (1982) Studies on the relation of γ-hydroxybutyric acid (GHB) to γ-aminobutyric acid (GABA). Biochem Pharmacol 31: 3917–3923

Snead OC, Furner R, Liu CC (1989) In vivo conversion of γ-aminobutyric acid and 1,4,butanediol to γ-hydroxybutyric acid in rat brain: studies using stable isotopes. Biochem Pharmacol 38: 4375–4380

Snead OC, Hechler V, Vergnes M, Marescaux C, Maitre M (1990) Increased γ-hydroxybutyric acid receptors in thalamus of a genetic animal model of petit mal epilepsy. Epilepsy Res 7: 121–128

Speth RC, Wastek GJ, Johnson PC, Yamamura HI (1978) Benzodiazepine binding in human brain; characterization using ^{3}H-flunitrazepam. Life Sci 22: 859–866

Steriade M, Llinas R (1988) The functional states of the thalamus and the associated neuronal interplay. Physiol Rev 68: 649–742

Tanaka T, Starke K (1980) Antagonist/agonist preferring alpha-adrenoceptors or alpha1/alpha2-adrenoceptors? Eur J Pharmacol 63: 191–194

Tran VT, Lebovitz R, Toll L, Snyder SH (1981) [^{3}H]Doxepine interactions with histamine H_1 receptors and other sites in guinea pig and rat brain homogenates. Eur J Pharmacol 70: 501–509

Vayer P, Mandel P, Maitre M (1987) Gamma-hydroxybutyrate, a possible neurotransmitter. Life Sci 41: 1547–1557

Vayer P, Ehrhardt J-D, Gobaille S, Mandel P, Maitre M (1988) Gamma-hydroxybutyrate distribution and turnover rates in discrete brain regions of the rat. Neurochem Int 12: 53–59

Vergnes M, Marescaux C, Micheletti G, Reis J, Depaulis A, Rumbach L, Warter JM (1982) Spontaneous paroxysmal electroclinical patterns in rat: a model of generalized non-convulsive epilepsy. Neurosci Lett 33: 97–101

Vergnes M, Marescaux C, Micheletti G, Depaulis A, Rumbach L, Warter JM (1984) Enhancement of spike and wave discharges by GABAmimetic drugs in rats with spontaneous petit-mal-like epilepsy. Neurosci Lett 44: 91–94

Vergnes M, Marescaux C, Depaulis A, Micheletti G, Warter JM (1987) Spontaneous spike and wave discharges in thalamus and cortex in a rat model of genetic petit mal-like seizures. Exp Neurol 96: 127–136

Vergnes M, Marescaux C, Depaulis A, Micheletti G, Warter JM (1990) Spontaneous spike-and-wave discharges in Wistar rats: a model of genetic generalized non-convulsive epilepsy. In: Avoli M, Gloor P, Kostopoulos G, Naquet R (eds) Generalized epilepsy: neurobiological approaches. Birkhäuser, Boston, pp 238–253

Winer BJ (1971) Statistical principles in experiment design. McGraw-Hill, New York, p 201

Wood P (1991) Pharmacology of the second messenger, cyclic guanosine 3',5'-monophosphate, in the cerebellum. Pharmacol Rev 43: 1–25
Yamamura HI, Snyder SH (1974) Muscarinic cholinergic binding in rat brain. Proc Natl Acad Sci 71: 1725–1729

Authors' address: Dr. R. Bernasconi, Research and Development Department, Pharmaceuticals Division, Ciba-Geigy Ltd., CH-4002 Basel, Switzerland

J Neural Transm (1992) [Suppl] 35: 179–188
© Springer-Verlag 1992

GABA$_B$ receptor antagonists: potential new anti-absence drugs

C. Marescaux[1], M. Vergnes[2], and R. Bernasconi[3]

[1] Clinique Neurologique, C.H.U. Strasbourg, [2] L.N.B.C., Centre de Neurochimie du CNRS, Strasbourg, France, and [3] Ciba-Geigy, Basel, Switzerland

Summary. The availability of new antagonists of the GABA$_B$ receptor which readily cross the blood-brain barrier has made it possible to investigate the role of GABA$_B$-receptor-mediated transmission in the control of spike-and-wave discharges (SWD) in a strain of rats (GAERS) with genetic absence epilepsy. Systemic administration of R-Baclofen, a GABA$_B$ agonist, increased the duration of SWD, or elicited SWD-like oscillations in the cortical EEG of non-epileptic control rats. Conversely, administration of CGP 35348, a GABA$_B$ antagonist, either i.p. or p.o., dose-dependently suppressed the spontaneous SWD, as well as the SWD aggravated by concomitant injection of various GABAmimetic drugs, GHB, or anticonvulsants known to exacerbate absence seizures. These results demonstrate the involvement of GABA$_B$-mediated neurotransmission in the development of SWD in generalized non-convulsive epilepsy. GABA$_B$ antagonists may thus be considered to be potentially specific anti-absence drugs.

Introduction

A strain of Wistar rats was selected in our laboratory for spontaneous occurrence of generalized non-convulsive seizures or absences (Genetic Absence Epilepsy Rats from Strasbourg, GAERS). The seizures are characterized by bilateral and synchronous spike-and-wave discharges (SWD) concomitant with a behavioural arrest. They are abolished by all the antiepileptics currently used in treatment of absence epilepsy, and they are exacerbated by the drugs which induce absence-like seizures in various experimental models. All the characteristics of these seizures agree with a model of generalized non-convulsive absence-type epilepsy. The usefulness of this model in predicting the pharmacological effects of new compounds has been repeatedly demonstrated (for review see Marescaux et al., this volume).

GABAergic transmission was shown to play a major role in the control of SWD: all GABA$_A$-mimetics dose-dependently increase the SWD in

GAERS as in other absence-like models (Fariello et al., 1980; King, 1979; Meldrum and Horton, 1980; Snead, 1990; Vergnes et al., 1984). We have previously noted that $GABA_B$ agonists also increased SWD (Vergnes et al., 1984). However, until recently, it was impossible to study the exact role of $GABA_B$ neuromediation, as no $GABA_B$ antagonist capable of crossing the blood-brain barrier efficiently was available.

In the present experiments in GAERS, we have examined the effects of systemically administered Baclofen (Hill and Bowery, 1981), a specific agonist, and CGP 35348 (Bittiger et al., 1990; Olpe et al., 1990), a specific antagonist of the $GABA_B$ receptor, which readily cross the blood-brain barrier.

Methods

Animals

Male adult rats from the GAERS strain, weighing 350–450 g, were fitted, under pentobarbitone anaesthesia (40 mg/kg i.p.) with 4 screw electrodes over the left and right frontoparietal cortex. Experiments were started after one week of recovery. Some experiments were also performed on non-epileptic rats from the control strain.

Procedure

Effects of R-Baclofen and CGP 35348 on spontaneous SWD in GAERS

R-Baclofen (1–8 mg/kg) and CGP 35348 (50–400 mg/kg), from Ciba-Geigy, Switzerland, were dissolved in 0.9% saline and administered intraperitoneally or orally in a volume of 2 ml/kg. Each treatment was given to a group of 6 rats. After 15 min habituation in the recording cage, the EEG was recorded for a 20-min reference period. The drugs were then administered i.p. or p.o. The solvent was given for the zero dose. EEG's were recorded immediately for 120 to 240 min.

Effects of CGP 35348 on pharmacologically aggravated SWD

The duration of spontaneous SWD was increased by various compounds: R-Baclofen (4 mg/kg), gamma-hydroxybutyrate (GHB, 375 mg/kg), THIP (8 mg/kg), SKF 89976 (30 mg/kg), gamma-vinyl-GABA (GVG, 600 mg/kg), carbamazepine (CBZ, 20 mg/kg), and phenytoin (PHT, 50 mg/kg). SKF 89976 was suspended in 0.9% NaCl with 2 drops of Tween 80/10 ml. CBZ was dissolved in Molecusol (Ciba-Geigy). The commercial solution of PHT was injected. All the other compounds were dissolved in 0.9% saline. The drugs were injected i.p. in a volume of 2 ml/kg, into groups of 6 animals.

Two different procedures were used: (1) the aggravating drug was administered 20 min before CGP 35348 (0, 200 and 400 mg/kg); (2) CGP 35348 was first injected 40 min before the injection of the aggravating drug. EEG's were recorded continuously from 20 min before the first injection until 80 to 120 min after the second.

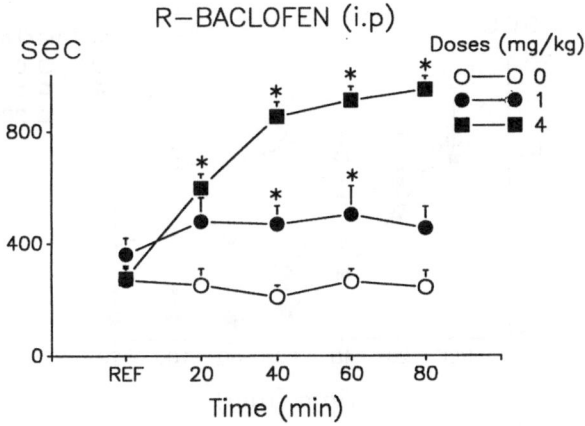

Fig. 1. Effects of i.p. administration of R-Baclofen at different doses in GAERS. Results are expressed as mean ± SEM cumulative duration of SWD in sec per 20-min period before (REF) and after injection. *p < 0.02 vs dose 0

As the effect of GVG on GABA accumulation is delayed, GVG was injected 3 hrs before CGP 35348 and the EEG recorded one hour later for 40 min.

Analysis of results

The results were analysed by consecutive 20-min periods. The total duration of SWD per 20-min period was measured for each animal and expressed as the mean cumulative duration of SWD per group of 6 rats. The post-injection periods were compared to the reference period using a non-parametric analysis of variance for related samples (Friedman test). Results for each dose versus the zero control dose were then compared using the Wilcoxon test.

Results

R-Baclofen induced a rapid and dose-dependent increase in the duration of SWD, which reached a peak between 40 and 60 min post-injection (Fig. 1). At doses above 4 mg/kg, the SWD became quite continuous, producing a status epilepticus. The effect of the highest doses of R-Baclofen lasted for about 240 min.

R-Baclofen was also administered to non-epileptic rats from the control strain in doses of 2, 4 and 8 mg/kg. At 4 and 8 mg/kg, it induced bilateral short paroxysmal discharges of irregular and slow (4–6 c/s) spikes and waves, while behavioural activity was interrupted (results not shown). In both GAERS and controls, the highest doses provoked sedation and myorelaxation.

CGP 35348, 50–400 mg/kg i.p., induced a dose-dependent and progressive suppression of SWD, with a peak effect 60 min after the injection.

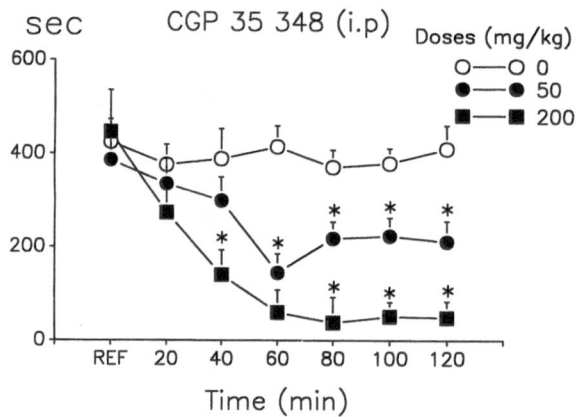

Fig. 2. Effects of i.p. administration of CGP 35348 at different doses in GAERS. Conventions as in Fig.1

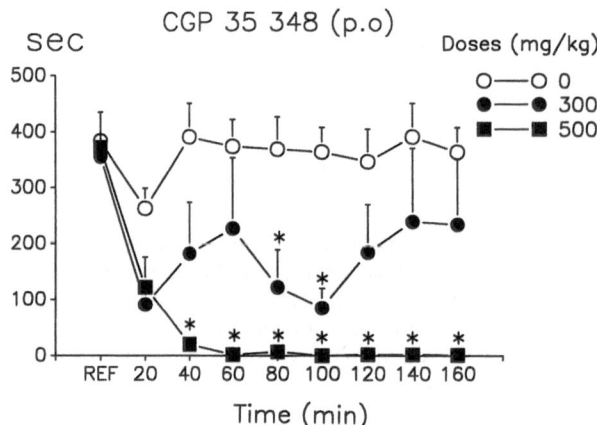

Fig. 3. Effects of oral administration of CGP 35348 at different doses in GAERS. Conventions as in Fig. 1

50% and 90% suppression were obtained with 50 and 200 mg/kg respectively (Fig. 2). This effect lasted for 180 to 240 min according to the dose. Oral administration of CGP 35348 suppressed SWD completely at the 500-mg/kg dose and by 50% at 300 mg/kg (Fig. 3). CGP 35348 produced no side-effects: the animals remained awake with a normal desynchronized background EEG.

An increase in the cumulative duration of SWD, by 150 to 300%, was obtained with the following drugs: R-Baclofen, a GABA$_B$ agonist, THIP, a GABA$_A$ agonist, SKF 89976, an inhibitor of GABA reuptake, GVG, an inhibitor of GABA transaminase, CBZ and PHT, both anticonvulsants that aggravate absence seizures, and GHB. In all cases, CGP 35348 dose-

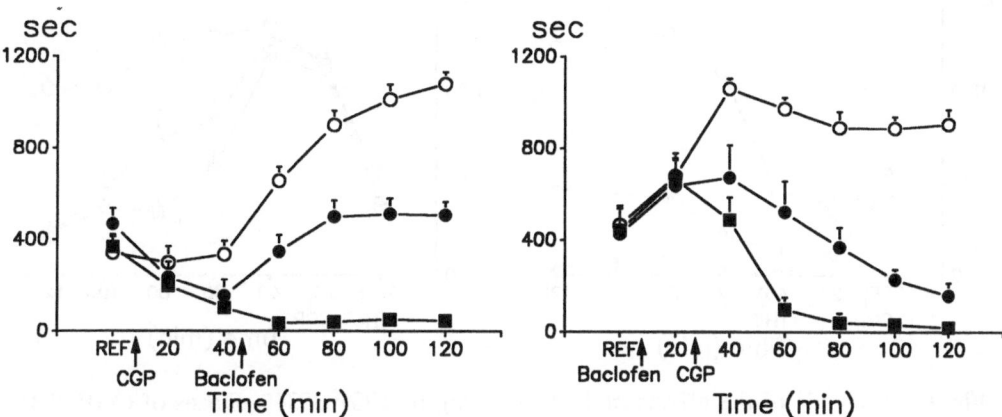

CGP 35348 + BACLOFEN

Fig. 4. Blockade of the effects of R-Baclofen 4 mg/kg by CGP 35348. Doses of CGP 35348: open circles, 0 mg/kg (control experiment); black circles, 200 mg/kg; black squares, 400 mg/kg. Left panel, CGP 35348 administered before, and right panel, CGP 35348 administered after R-Baclofen. Conventions as in Fig. 1

CGP 35348 + GHB

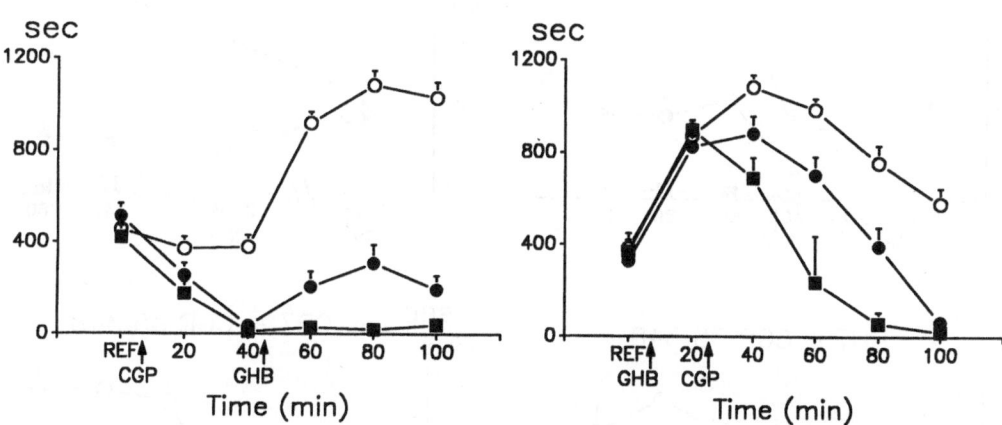

Fig. 5. Blockade of the effects of GHB 375 mg/kg by CGP 35348. Doses of CGP 35348: open circles, 0 mg/kg (control experiment); black circles, 200 mg/kg; black squares, 400 mg/kg. Left panel, CGP 35348 administered before, and right panel, CGP 35348 administered after GHB. Conventions as in Fig. 1

dependently suppressed the SWD when they were aggravated by one of the previous treatments, or prevented the aggravation if it was injected before the aggravating drug. The suppressant effect of CGP 35348 was significant at 200 mg/kg. CGP 35348 completely abolished the SWD at 400 mg/kg, except after THIP, in which case complete suppression was obtained with 600 mg/kg (Fig. 4, 5, 6, 7).

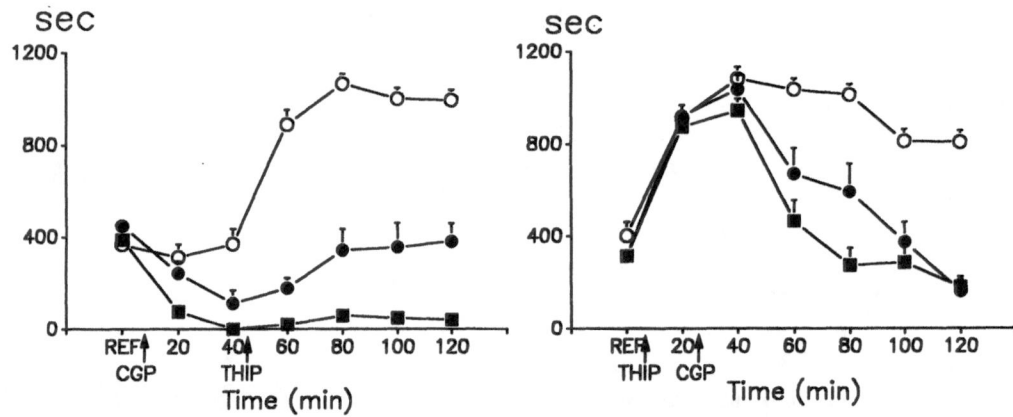

Fig. 6. Blockade of the effects of THIP 8 mg/kg by CGP 35348. Doses of CGP 35348: open circles, 0 mg/kg (control experiment); black circles, 400 mg/kg; black squares, 600 mg/kg. Left panel, CGP 35348 administered before, and right panel, CGP 35348 administered after THIP. Conventions as in Fig. 1

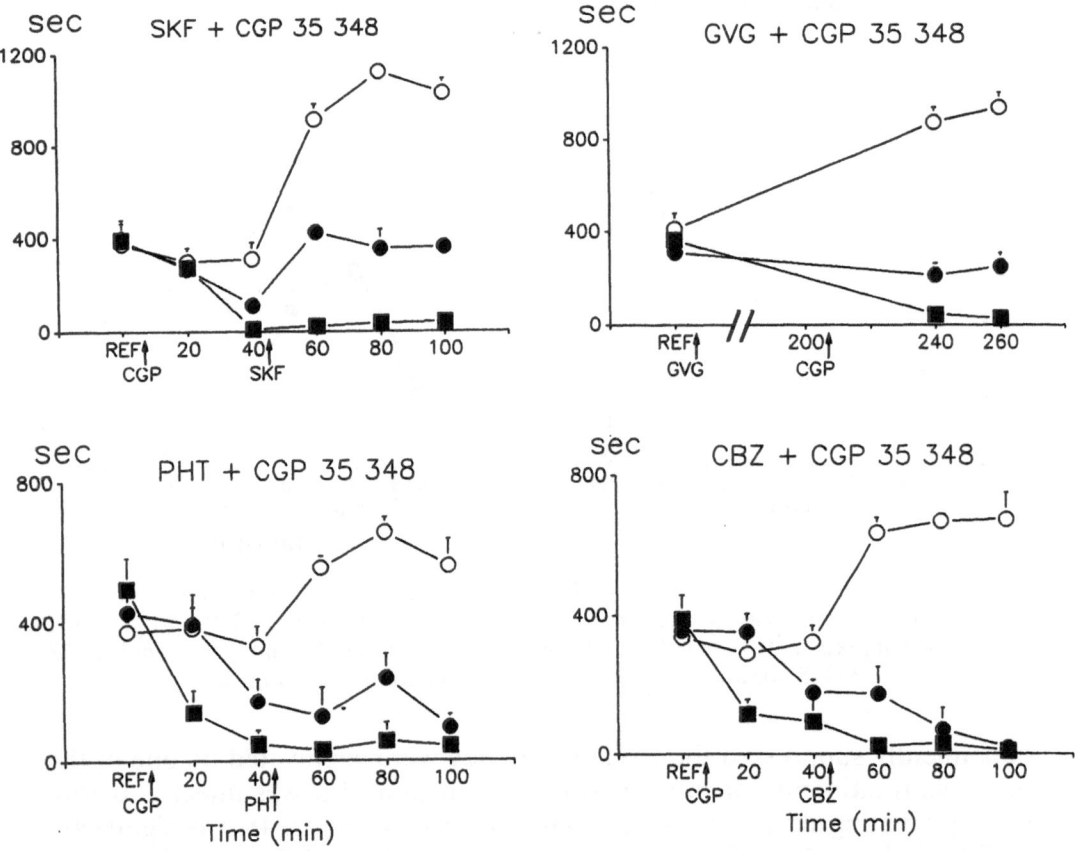

Fig. 7. Blockade of the effects of SKF 89976 20 mg/kg, GVG 600 mg/kg, PHT 50 mg/kg and CBZ 20 mg/kg by pretreatment with CGP 35348. Doses of CGP 35348: open circles, 0 mg/kg (control experiment); black circles, 200 mg/kg; black squares, 400 mg/kg. Conventions as in Fig. 1

Discussion

Previous experiments have shown that the absence-epilepsy seizures of
GAERS involved a thalamocortical substrate, the SWD being possibly
initiated in the ventrolateral thalamus (Vergnes and Marescaux, this
volume). Whereas most of the neurotransmitters participate in the control
of SWD, GABA appears to play a prominent role (Marescaux et al.,
this volume). All the GABAmimetics tested (muscimol and THIP,
GABA$_A$ agonists; GVG and L-cycloserine, inhibitors of GABA transaminase, SKF 89976, an inhibitor of GABA reuptake) increased the frequency and the duration of SWD (Vergnes et al., 1984; Marescaux et al.,
this volume). Similar results were obtained in all models of absence
epilepsy in animals and also in human pathology (Fariello et al., 1980;
Fariello and Golden, 1987; King, 1979; Meldrum and Horton, 1980; Snead,
1990). These data suggest that petit mal absences constitute a particular
form of epilepsy which may be related to an excess of GABAergic inhibition, in contradistinction to partial or generalized convulsive epilepsies
(Fariello and Golden, 1987; Fromm and Kohli, 1972; Fromm and Terrence,
1987; Gloor and Fariello, 1988).

The present data suggest that the activity of GABA$_B$ receptors is critical
in the generation of SWD. Baclofen not only increases SWD in epileptic
rats, but also induces paroxysmal discharges of oscillatory activity similar to
spikes and waves in control rats free of spontaneous SWD. By contrast, i.p.
or p.o. administration of CGP 35348 efficiently suppresses the SWD without
any apparent side-effects.

This prominent-role of GABA$_B$ neurotransmission in the induction of
SWD is confirmed by the fact that the GABA$_B$ antagonist suppresses or
prevents SWD aggravated by GABA$_A$ or GABA$_B$ agonists, by inhibitors of
GABA reuptake or transamination, by GHB or by some anticonvulsants
(CBZ, PHT). Similarly, CGP 35348 also suppressed SWD induced in
non-epileptic rats by GHB or pentylenetetrazole (Snead, personal
communication).

Similar results were obtained after intrathalamic micro-injections of a
GABA$_B$ agonist and antagonist. R-Baclofen injected into the specific relay
nuclei or the reticular nucleus of the thalamus increased the duration of
SWD in GAERS and elicited rhythmic oscillations on the cortical EEG in
non-epileptic control rats. By contrast, intrathalamic injections of CGP
35348 into the same nuclei suppressed the spontaneous SWD or the SWD
aggravated by prior injection of THIP, muscimol, GVG, SKF 89976 or
GHB (Liu et al., 1992 and unpublished results from our laboratory).

Thalamic neurons that are reciprocally connected with their cortical
projections appear to be predominantly involved in the generation of
rhythmic activities and most probably of SWD in absence seizures (Gloor,
1988; Jones, 1988; Steriade and Deschenes, 1984; Steriade et al., 1987). The
capacity of thalamic relay neurons to generate repeated bursts of action
potentials is related to the occurrence of rhythmic low-threshold calcium

currents (LTCC), which are deinactivated when the cell membrane is hyper-polarized. These oscillations would be controlled by the GABAergic afferents from the reticular neurons, which synchronize the thalamic activity through their widespread intrathalamic connections (Llinás and Geijo-Barrientos, 1988; Coulter et al., 1990; Jones, 1988).

$GABA_B$-receptor activation mediates a late and long-lasting inhibitory post-synaptic potential (IPSP), which produces the hyperpolarization necessary for LTCC to be elicited (Crunelli and Leresche, 1991). The LTCC were shown to underlie episodes of sleep spindles, and possibly also of SWD (Coulter et al., 1990; Gloor, 1988).

The suppression of SWD obtained by systemic as well as by intra-thalamic administration of $GABA_B$ antagonists is in accord with these cellular mechanisms and suggests that the $GABA_B$-mediated late IPSP in thalamic neurons is critical for oscillatory activity, and hence for the development of SWD. Neither spontaneous, nor pharmacologically aggravated or evoked SWD are recorded after blockade of the $GABA_B$ receptors.

Whereas SWD are also aggravated by systemic or intrathalamic administration of $GABA_A$ agonists, blockade of $GABA_A$ receptors did not reduce SWD at non-convulsive doses, suggesting that the early $GABA_A$-mediated IPSP is less critical in development of SWD. However, interference with convulsive effects may occlude such a mechanism (Liu et al., 1991).

Our results suggest that a possible dysfunction in $GABA_B$ receptor-mediated transmission may underlie episodes of genetically determined SWD. Therefore, the density and affinity of $GABA_B$ receptors was investigated in GAERS (Knight and Bowery, this volume). The absence of abnormality in these characteristics does not exclude the possibility of a modified transduction of these receptors to the second messanger in GAERS.

Another implication concerns the therapeutic use of $GABA_B$ antagonists as potential new anti-absence drugs. This anti-absence effect depends on hitherto unknown mechanisms, as the anti-epileptics commonly used in absence epilepsy, ethosuximide and valproate, do not interact with the $GABA_B$ receptor. CGP 35348 or other $GABA_B$ antagonists are selectively effective in absence epilepsy, as they do not reduce convulsant seizures (Karlsson et al., 1990).

In conclusion, the present data demonstrate that the $GABA_B$ receptor is involved in SWD in GAERS. The $GABA_B$ antagonists may prove to be of value in studies investigating the mechanisms involved in the generation of SWD in absence-epilepsy. Moreover, they are potential new anti-absence drugs.

Acknowledgements

Special thanks are given to A. Boehrer for technical assistance. This work was supported by grants from INSERM (CAR n° 400019) and from "La Fondation pour la Recherche Médicale".

References

Bittiger H, Froestl W, Hall R, Karlsson G, Klebs K, Olpe HR, Pozza MF, Steinmann MW, Van Riezen H (1990) Biochemistry, electrophysiology and pharmacology of a new GABA$_B$ antagonist: CGP 35348. In: Bowery NG, Bittiger H, Olpe HR (eds) GABA$_B$ receptors in mammalian function. Wiley, Chichester, pp 47–80

Coulter DA, Huguenard JR, Prince DA (1990) Cellular actions of petit mal anticonvulsants: implication of thalamic low-threshold calcium current in generation of spike-wave discharge. In: Avoli M, Gloor P, Kostopoulos P, Naquet R (eds) Generalized epilepsy: neurobiological approaches. Birkhäuser, Boston, pp 425–435

Crunelli V, Leresche N (1991) A role for GABA$_B$ receptor in excitation and inhibition of thalamocortical cells. Trends Neurosci 14: 16–21

Fariello RG, Golden GT (1987) The THIP-induced model of bilateral synchronous spike and wave in rodents. Neuropharmacology 26: 161–165

Fariello RG, Golden GT, Black JA (1980) Potentiation of a feline model of corticoreticular epilepsy by systematically administered inhibitory amino acids. In: Canger R, Angeleri F, Penry JK (eds) Advances in Epileptology, XIth Epilepsy International Symposium. Raven Press, New York, pp 339–342

Fromm GH, Kohli CM (1972) The role of inhibitory pathways in petit mal epilepsy. Neurology 22: 1012–1020

Fromm GH, Terrence CF (1987) Effect of antiepileptic drugs on the brainstem. In: Fromm GH, Faingold CL, Browning RA, Burnham WM (eds) Epilepsy and the reticular formation: the role of the reticular core in convulsive seizures. Alan R Liss, New York, pp 119–136

Gloor P (1988) Neurophysiological mechanism of generalized spike-and-wave discharges and its implication for understanding absence seizures. In: Myslobodsky MS, Mirsky AF (eds) Elements of petit mal epilepsy. Peter Lang, New York, pp 159–209

Gloor P, Fariello RG (1988) Generalized epilepsy: some of its cellular mechanisms differ from those of focal epilepsy. TINS 11: 63–68

Hill DR, Bowery NG (1981) (^{3}H)baclofen and (^{3}H)GABA bind to bicuculline-insensitive GABA$_B$ sites in rat brain. Nature 290: 149–152

Jones EG (1988) Modern views of cellular thalamic mechanisms. In: Bentivoglio M, Spreafico R (eds) Cellular thalamic mechanisms. Elsevier, Amsterdam, pp 1–22

Karlsson G, Schmutz M, Kolb C, Bittiger H, Olpe HR (1990) GABA$_B$ receptors and experimental models of epilepsy. In: Bowery NG, Bittiger H, Olpe HR (eds) GABA$_B$ receptors ion mammalian function. Wiley, Chichester, pp 349–365

King GA (1979) Effects of systematically applied GABA agonists and antagonists on wave-spike ECoG activity in rat. Neuropharmacology 18: 47–55

Knight AR, Bowery NG (1992) GABA receptors in rats with spontaneous generalized nonconvulsive epilepsy (this volume)

Liu Z, Vergnes M, Depaulis A, Marescaux C (1991) Evidence for a critical role of GABAergic transmission within the thalamus in the genesis and control of absence seizures in the rat. Brain Res 545: 1–7

Liu Z, Vergnes M, Depaulis A, Marescaux C (1992) Involvement of intrathalamic GABA$_B$ neurotransmission in the control of absence seizures in the rat. Neuroscience (in press)

Llinás RR, Geijo-Barrientos E (1988) In vitro studies of mammalian thalamic and reticularis thalami neurons. In: Bentivoglio M, Spreafico R (eds) Cellular thalamic mechanisms. Elsevier, Amsterdam, pp 23–33

Marescaux C, Vergnes M, Depaulis A (1992) Genetic absence epilepsy in rats from Strasbourg. A review (this volume)

Meldrum B, Horton R (1980) Effects of the bicyclic GABA agonist, THIP, on myoclonic and seizure responses in mice and baboons with reflex epilepsy. Eur J Pharmacol 61: 231–237

Olpe H, Karlsson G, Pozza MF, Brugger F, Steinmann M, Riezen HV, Fagg G, Hall RG, Froestl W, Bittiger H (1990) CGP 35348: a centrally active blocker of GABA$_B$ receptors. Eur J Pharmacol 187: 27–38

Snead OC (1990) The ontogeny of GABAergic enhancement of the gamma-hydroxybutyrate model of generalized absence seizure. Epilepsia 31: 363–368

Steriade M, Deschenes M (1984) The thalamus as a neuronal oscillator. Brain Res Rev 8: 1–63

Steriade M, Domich L, Oakson G, Deschenes M (1987) The deafferented reticular thalamic nucleus generates spindle rhythmicity. J Neurophysiol 57: 260–273

Vergnes M, Marescaux C (1992) Cortical and thalamic lesions in rats with genetic absence epilepsy (this volume)

Vergnes M, Marescaux C, Micheletti G, Depaulis A, Rumbach L, Water JM (1984) Enhancement of spike and wave discharges by GABAmimetic drugs in rats with spontaneous petit mal-like epilepsy. Neurosci Lett 44: 91–94

Authors' address: Dr. C. Marescaux, Clinique Neurologique, Hôpital Civil, 1 place de l'Hôpital, F-67091 Strasbourg Cedex, France

J Neural Transm (1992) [Suppl] 35: 189–196

GABA receptors in rats with spontaneous generalized nonconvulsive epilepsy

A. R. Knight and **N. G. Bowery**

Department of Pharmacology, School of Pharmacy, London, United Kingdom

Summary. We have used the technique of autoradiography to study the binding of $[^3H]$-GABA to $GABA_A$ and $GABA_B$ receptors in brains taken from rats that are genetically predisposed to petit mal type seizures. A range of concentrations of $[^3H]$-GABA were employed to test the hypothesis that this predisposition was due to regional changes in either the number of $GABA_A$ or $GABA_B$ receptors, or affinity of GABA for these receptors.

We found no statistical difference in the levels of radioligand binding to $GABA_A$ and $GABA_B$ receptors in animals susceptible to seizures compared to control animals in any of the brain regions studied over the concentration range 25 nM to 400 nM. This indicated that there was no change in either the Kd (affinity) or Bmax (receptor number) in these animals and that the pharmacological basis for the efficacy of $GABA_B$ antagonists in this seizure condition probably lies elsewhere.

Introduction

The neurotransmitter GABA (γ-amino-butyric acid) exerts its influence via a number of receptors that can be distinguished pharmacologically and by their cellular location. Of the two pharmacologically distinguishable receptors, $GABA_A$ receptors have been described in the most detail and are multi-subunit, agonist gated ion channels that are uniquely sensitive to bicuculline (see Curtis et al., 1970). Activation of post-synaptic $GABA_A$ receptors results in an increase in chloride conductance across the cell membrane, the resulting hyperpolarisation suppressing neuronal firing.

GABA also activates a G protein linked receptor (Morishita et al., 1990) which has been designated $GABA_B$ (Hill and Bowery, 1981). Baclofen is a selective agonist for this receptor (Bowery et al., 1983) which is present at both pre- and post-synaptic sites. Stimulation of post-synaptic $GABA_B$ receptors results in an increase in K^+ efflux, producing hyperpolarization, and therefore inhibition of neuronal firing (Dutar and Nicoll, 1988). Activation of pre-synaptic $GABA_B$ receptors also suppresses neurotransmission by inhibiting synaptic release (Bowery et al., 1980) and it seems likely that this effect is mediated via an inhibition of Ca^{2+} influx and/or K^+ efflux. $GABA_B$ receptors are also able to influence intracellular events via the cAMP second messenger system (Wojcik and Neff, 1983), although an exact physiological role for this coupling has yet to be ascribed.

The hypothesis that $GABA_A$ receptors play a role in the initiation and control of seizure activity is supported by a variety of data. For instance $GABA_A$ receptor agonists such as muscimol (Enna and Beutler, 1985), or compounds that increase GABA concentration such as γ-vinyl GABA (Meldrum and Horton, 1978) or which enhance $GABA_A$ receptor function such as benzodiazepines (Meldrum and Braestrup, 1984), are anticonvulsant whilst classical $GABA_A$ antagonists such as picrotoxin and bicuculline are pro-convulsant (Enna and Beutler, 1985). The concentration of GABA is elevated in the cerebrospinal fluid in patients with epilepsy (Wood et al., 1979) which might suggest the involvement of a feedback mechanism since this implies enhanced GABA-ergic function in these patients.

$GABA_B$ receptor function has also been implicated in the generation of seizure activity. Baclofen has been shown to be pro-convulsant in models of epilepsy (Swartzwelder et al., 1987; Cotterell and Robertson, 1987). GABA autoreceptors appear to be of the $GABA_B$ class and their activation suppresses the release of GABA which may produce net neuronal excitation due to a reduction in inhibitory input. In addition Crunelli and Leresche (1991) have recently suggested that $GABA_B$ receptor activation in thalamic neurones, which produces a late prolonged ipsp, may allow the reactivation of the T-current for Ca^{2+}. This could facilitate the generation of spike activity comparable to that occurring in petit mal type seizures.

Vergnes et al. (1984, 1987) have recently described a strain of rats that may represent an animal model of petit mal type epilepsy. These rats are genetically susceptible to spontaneous behavioural arrests which occur simultaneously with electroencephalographic spike and wave discharges. These symptoms represent a condition known as spontaneous generalized non convulsant epilepsy and are believed to represent a model of non-convulsive or petit mal epilepsy in humans. These animals are sensitive to a variety of drugs that interact with the GABA receptors. GABA mimetics, including agonists for both $GABA_A$ and $GABA_B$ receptors, and compounds such as γ-vinyl GABA that increase the concentration of endogenous GABA, all increase the number and duration of spike and wave discharges (Vergnes et al., 1984). The knowledge that spike and wave discharges are sensitive to pharmacological manipulation by drugs that interact with GABA-ergic systems and the apparent intimate relationship between GABA and epilepsy leads to the hypothesis that the basis of the genetic susceptibility of these rats might lie in an alteration of GABA transmission. Whilst $GABA_A$ antagonists do not alter the slow wave discharges (Vergnes et al., 1984), recent studies by Marescaux et al. (1992) indicate that $GABA_B$ receptor antagonists can reduce the seizure discharges in this animal model. We have therefore investigated the affinity and density of GABA receptor binding sites in a variety of brain regions and focussing particular interest on the $GABA_B$ receptor in this model using the technique of quantitative autoradiography.

Materials and methods

The brains of 3 Wistar rats that showed spontaneous spike and wave discharges in their EEG and the brains of 3 animals from the same strain with normal EEGs were kindly obtained from Dr. C. Marescaux. These brains were stored at $-70°C$ prior to assay. $10\,\mu m$ thick parasagittal sections were cut from each brain and thaw mounted onto clean microscope slides. These sections were stored at $-20°C$ for at least 18 hours. Radioligand binding to the slide mounted sections was carried out according to the method of Bowery et al. (1987). Briefly, sections were thawed at room temperature for 1 hour prior to incubation in assay buffer (with or without $CaCl_2$) for 40 minutes. The slides were subsequently dried in air. GABA binding sites were labelled in triplicate sections from each of the 6 brains. Non specific binding was determined in triplicate in consecutive sections. The assay was performed by pipetting onto the slide $100\,\mu l$ of assay buffer containing one of 6 concentrations of [^{3}H]-GABA (90.1 Ci/mmol, NEN) in the range $25{-}400\,nM$. $GABA_B$ binding sites were labelled selectively in the presence of $40\,\mu M$ isoguvacine and $2.5\,mM$ $CaCl_2$. Non-specific binding was determined with $100\,\mu M$ $(-)$baclofen (Ciba-Geigy). $GABA_A$ binding sites were labelled selectively in the absence of calcium ions and in the presence of $100\,\mu M$ $(-)$baclofen. Non-specific binding was determined with $100\,\mu M$ isoguvacine. The binding reaction was allowed to reach equilibrium (20 min) after which the incubation medium was aspirated from the slide. Each slide was rinsed twice for 3 seconds in ice cold assay buffer followed by a rapid rinse in ice cold water. The sections were dried in a stream of air.

Bound radioactivity was visualized by placing the sections in contact with Hyperfilm (LKB) for 1 to 3 weeks. The latent images produced by the action of tritium on the emulsion were developed in D-19 developer (Kodak) and fixed in Kodak Unifix.

The optical densities of discrete areas of each autoradiogram were measured using a Quantimet 970 image analyzer. The quantity of bound radioactivity in the sections was estimated by comparison with polymer standards (Amersham). Areas of the autoradiogram corresponding to specific brain nuclei were identified using a stereotaxic atlas of the rat brain (Paxinos and Watson, 1986).

Results

The distribution of GABA binding to $GABA_B$ receptors in parasagittal sections of epileptic and non epileptic rat brain is shown in Fig. 1. The distribution patterns of $GABA_B$ binding sites in the brains of animals susceptible and not susceptible to seizures were the same. The amount of bound radioactivity was quantified for each of the radioligand concentrations used in selected brain areas, which included: the inner cortex (laminae V–VI), outer cortex (laminae I–IV), hippocampus CA1 oriens, lateral dorsal thalamic nucleus, latero-posterior thalamic nucleus, ventral posterior thalamic nucleus, striatum, the molecular layer and granule cell layer of the cerebellum. These data are shown in Fig. 2. There was no significant difference overall in the levels of specific binding (ANOVA, $F_{(1,4)} = 0.00$, P

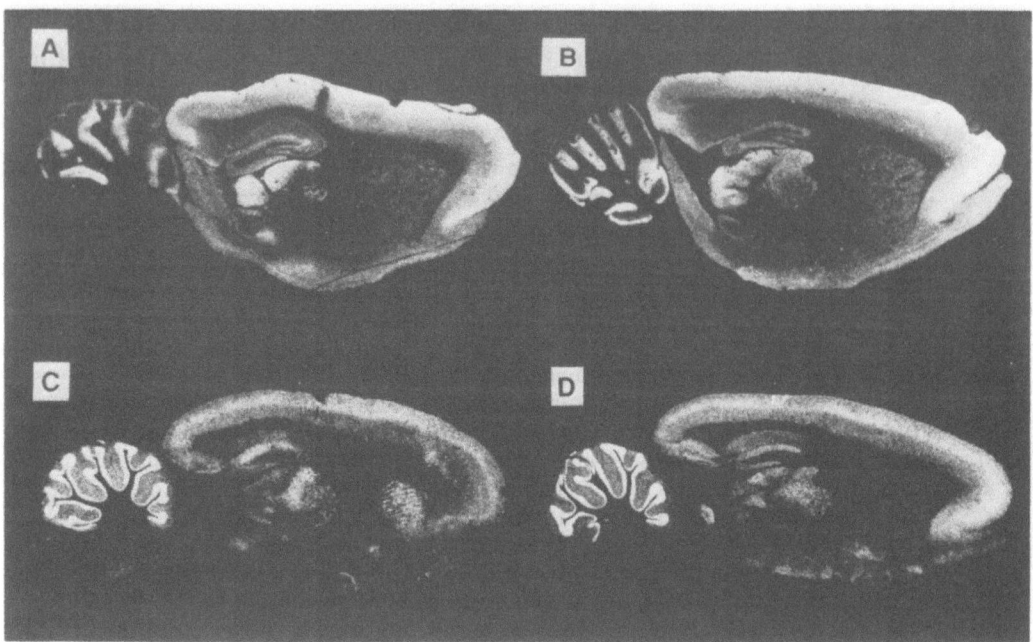

Fig. 1. Autoradiographs showing the regional distribution of GABA$_B$ binding sites in parasagittal sections approximately 1 mm from midline in (**A**) epileptic and (**B**) non-epileptic rat brain, and GABA$_A$ binding sites in sections from (**C**) epileptic and (**D**) non-epileptic rat brain. Light areas depict regions of high densities of binding. Experiments were performed using 50 nM [^{3}H]-GABA, under the conditions described in the text

= 0.986), neither was there any difference in any of the areas selected for study ($F_{(1,4)} < 0.59$, $p > 0.48$).

Data are also represented in Fig. 2 for [^{3}H]-GABA binding to GABA$_A$ receptors. There was no significant difference in the overall levels of binding to the GABA$_A$ receptor in these animals compared to controls (ANOVA, $F_{(1,4)} < 0.10$, $p > 0.764$), nor was there any significant difference in any of the specific nuclei ($F_{(1,4)} < 1.29$, $p > 0.32$).

Saturation analysis was performed on the binding data to determine the Kd of [^{3}H]-GABA for the GABA$_B$ receptor in epileptic and non epileptic brains. The mean Kd for [^{3}H]-GABA in epileptic brains was 64.48 ± 9.01 nM (mean ± S.E.M., n = 7) which did not differ significantly for the Kd obtained in control brains (56.92 ± 10.079 nM, n = 7, p > 0.10 Student "t" test).

Discussion

Comparison of the images generated by autoradiography for GABA$_B$ receptors bore a close resemblance to those already published (Bowery et

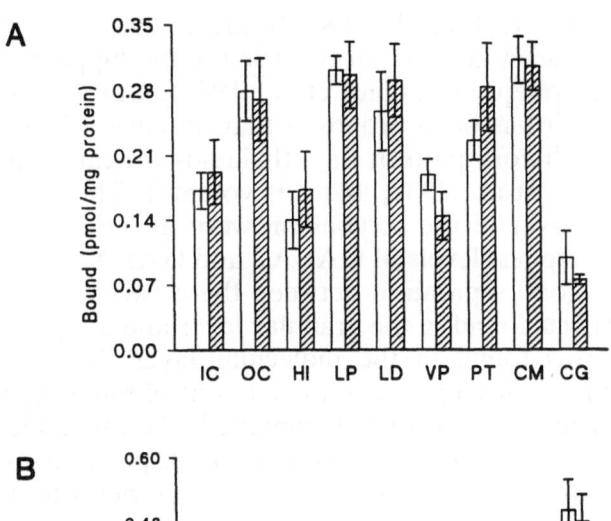

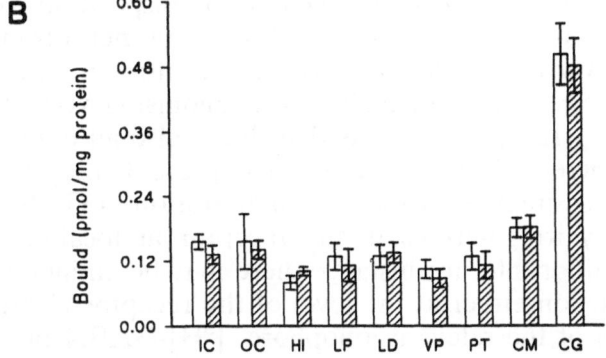

Fig. 2. Density of [³H]-GABA binding to (**A**) GABA$_B$ and (**B**) GABA$_A$ sites in specific nuclei in sections from epileptic rats (shaded bars) and control rats (open bars). The brain areas measured were inner cortex (IC), outer cortex (OC), Hippocampus CA1 oriens (H$_1$), Latero-posterior thalamic nucleus (LP), Lateral dorsal thalamic nucleus (LD), Ventral posterior thalamic nucleus (VP), Posterior thalamic muclear group (PT) and the molecular layer (CM) and granule cell layer (CG) of the cerebellum. Values represent optical density measurements in 3 sections from each of 3 individual rat brains (n = 3). Optical densities were converted to concentrations by comparison with polymer standards. The binding assay was performed with 50 nM [³H]-GABA as described in the text

al., 1987; Chu et al., 1990) with high levels of binding in the cerebellum granule cell layer, the zonal layer of the superior colliculus and in thalamus. A comparison of the images from epileptic animals and those for control animals revealed no gross structural changes, so quantal analysis of the optical density of areas of the film representing specific brain nuclei was undertaken. Some of the areas selected for study have been implicated in epileptic events (see below), whilst the others have high densities of GABA$_B$ binding and are therefore likely to be areas where GABA$_B$ function is important.

The first class of structures is represented by cortex, hippocampus and the thalamic nuclei. Hippocampus has frequently been used in in vitro models of epilepsy (Ault et al., 1983; Swartzwelder et al., 1986). However spike and wave discharges are not recorded in the hippocampus in rats with genetic absence epilepsy (Vergnes et al., 1987). Cortex and thalamus were selected because it has been proposed that absence seizures are caused by an abnormal oscillatory pattern in a thalamo-cortical loop (McLachlan et al., 1984; Vergnes et al., 1987, and this volume). The molecular layer and the granule cell layer of the cerebellum were included as two regions that represented the highest levels of $GABA_A$ and $GABA_B$ binding in the brain, and provided a good separation between these two sites, a high ratio of A to B sites (5:1) was found in the granular layer and a high ratio of B sites to A sites (2:1) was found in the molecular layer. None of these regions showed a significant discrepancy in the amount of binding present in seizure prone animals compared to control animals. It therefore seems unlikely that the physiological mediator of the genetic predisposition to non-convulsive epilepsy is due to an alteration in GABA-ergic neurotransmission caused either by an over abundance of $GABA_B$ receptors or by an increase in the affinity of these receptors for their natural agonist GABA. Data shown here for the $GABA_A$ receptor suggests that this is also an unlikely candidate for the cause of increased seizure events. The present study however, does not preclude the possibility that there is an alteration in GABA-ergic transmission caused by irregularities in the receptor at localities other than the agonist recognition site, for instance there may be an increased efficiency in signal transduction either at the level of the receptor-G protein interaction or at the level of the relevant ionophore. [^{3}H]-GABA binding to $GABA_A$ receptors is reduced in convulsive epilepsy, both in humans (Lloyd et al., 1981) and in some animal models such as DBA/2 mice which are susceptible to audiogenic seizures (Horton et al., 1982). Our study therefore highlights a possible dichotomy in the pathology of convulsive and non convulsive epilepsy. Baclofen exacerbates spike and wave discharges in these animals (Vergnes et al., 1984) whilst the specific $GABA_B$ antagonist CGP35348, has been shown to inhibit them, suggesting that drugs which interact the $GABA_B$ receptor may be of value in treating petit mal epilepsy (Marescaux et al., this volume). By contrast the value of $GABA_B$ antagonists as therapeutic agents for grand mal epilepsy is uncertain (Karlsson et al., 1990). Baclofen has been shown to be either pro-convulsant (Cotterell and Robertson, 1987) or anticonvulsant (Menon and Vivonia, 1981). These discrepancies may be due in part to the relative importance of post-synaptic inhibition and presynaptic disinhibition in the control of seizures. The complex relationship between inhibition and disinhibition is further emphasised in electrophysiological studies, some of which suggest that baclofen inhibits epileptiform activity (Swartzwelder et al., 1986), whilst others suggest that ictal (seizure) events are independent of $GABA_B$ receptor control but inter-ictal (between seizure) events are inhibited by baclofen (Swartzwelder et al., 1987). Since a decrease in inter-ictal events increases ictal activity, seizures and convulsions might be more readily expressed in the presence of a $GABA_B$ agonist. The determination of the relative importance of post-synaptic inhibition and pre-synaptic disinhibition in the role of $GABA_B$ receptors in epilepsy awaits the development of specific drugs for receptors at these different cellular locations.

In conclusion, the present study suggests that altered GABA receptor expression is unlikely to be responsible for the mechanism underlying spontaneous generalized non-convulsive epilepsy in rats. However the results do not exclude the possibility that a change in receptor activation or the response to receptor activation is associated with this seizure condition.

References

Ault B, Nadler JV (1983) Anticonvulsant like actions of baclofen in the rat hippocampal slice. Br J Pharmacol 78: 701–708

Bowery NG, Hill DR, Hudson AL, Doble A, Middlemiss DN, Shaw J, Turnbull M (1980) (−)Baclofen decreases neurotransmitter release in the mammalian CNS by an action at a novel GABA receptor. Nature 283: 92–94

Bowery NG, Hill DR, Hudson AL (1983) Characteristics of $GABA_B$ receptor binding sites on rat whole brain synaptic membranes. Br J Pharmacol 78: 191–206

Bowery NG, Hudson AL, Price GW (1987) $GABA_A$ and $GABA_B$ receptor site distribution in the rat central nervous sytem. Neuroscience 20: 365–383

Chu DCM, Albin RL, Young AB, Penney JB (1990) Distribution and kinetics of $GABA_B$ binding sites in rat central nervous system: a quantitative autoradiographic study. Neuroscience 34: 341–357

Cotterell GA, Robertson HA (1987) Baclofen exacerbates epileptic myoclonus in kindled rats. Neuropharmacology 26: 645–648

Crunelli V, Leresche N (1991) A role for $GABA_B$ receptors in excitation and inhibition of thalamocortical cells. Trends Neurosci 14: 16–21

Curtis DR, Duggan AW, Felix D, Johnston GAR (1970) GABA, bicuculline and central inhibition. Nature 226: 1222–1224

Dutar P, Nicoll RA (1988) A physiological role for $GABA_B$ receptors in the central nervous system. Nature 332: 156–158

Enna SJ, Beutler JA (1985) GABA receptor as a site for antiepileptic drug action In: Bartholini G, Bossi L, Lloyd KG, Morselli PL (eds) Epilepsy and GABA receptor agonists: basic and therapeutic research. Raven Press, New York, pp 195–201

Hill DR, Bowery NG (1981) ^{3}H-Baclofen and ^{3}H-GABA bind to bicuculline insensitive $GABA_B$ sites in rat brain. Nature 290: 149–152

Horton RW, Prestwich SA, Meldrum BS (1982) γ-Aminobutyric acid and benzodiazepine binding sites in audiogenic seizure susceptible mice. J Neurochem 39: 864–870

Karlsson G, Schmutz M, Kolb C, Bittiger H, Olpe H-R (1990) $GABA_B$ receptors and experimental models of epilepsy In: Bowery NG, Bittiger H, Olpe H-R (eds) $GABA_B$ receptors in mammalian function. John Wiley, Chichester, pp 349–365

Lloyd KG, Munari C, Bossi L, Stoeffels C, Talairach J, Morselli PL (1981) Biochemical evidence for the alterations of GABA-mediated synaptic transmission in pathological brain tissue (stereo EEG or morphological definition) for epileptic patients. In: Morselli PL, Lloyd KG, Loscher W, Meldrum B, Reynolds EH (eds) Neurotransmitters, seizures and epilepsy. Raven Press, New York, pp 325–338

Marescaux C, Vergnes M, Bernasconi R (1992) $GABA_B$ receptor antagonists: potential new anti-absence drugs (this volume)

McLachlan RS, Gloor P, Avoli M (1984) Differential participation of some "specific" and "non-specific" thalamic nuclei in generalized spike and wave discharges of feline generalized penicillin epilepsy. Brain Res 307: 277–287

Meldrum B, Horton R (1978) Blockade of epileptic responses in the photosensitive baboon Papio papio by 2 irreversible inhibitors of GABA-transaminase, γ-acetylenic GABA (4-amino-hex-5-ynoic acid) and γ-vinyl GABA (4-amino-hex-5-enkoic acid). Psychopharmacology 5: 47–50

Meldrum B, Braestrup C (1984) GABA and the anticonvulsant action of benzodiazepines and related drugs. In: Bowery NG (ed) Actions and interactions of GABA and benzodiazepines. Raven Press, New York, pp 133–153

Menon MK, Vivonia CA (1981) Serotonergic drugs, benzodiazepines and baclofen block muscimol induced myoclonic jerks in a strain of mice. Eur J Pharmacol 73: 155–161

Micheletti G, Marescaux C, Vergnes M, Rumbach L, Warter JM (1985) Effects of GABA mimetics and GABA antagonists on spontaneous nonconvulsive seizures in Wistar rats. In: Bartholini G, et al (eds) L.E.R.S., vol 3. Raven Press, New York, pp 129–137

Morishita R, Kato K, Asano T (1990) $GABA_B$ receptors couple to G proteins Go, Go* and Gil but not Gi2. F.E.B.S. 271: 231–235

Paxinos G, Watson C (1986) The rat brain in stereotaxic coordinates. Academic Press, London

Swartzwelder HS, Bragdon AC, Sutch CP, Ault V, Wilson WA (1986) Baclofen suppresses hippocampus epileptiform activity at low concentrations without suppressing synaptic transmission. J Pharmacol Exp Ther 237: 881–887

Swartzwelder HS, Lewis DV, Anderson WW, Wilson WA (1987) Seizure-like events in brain slices: suppression by interictal activity. Brain Res 410: 362–366

Vergnes M, Marescaux C (1992) Cortical and thalamic lesions in rats with genetic absence epilepsy (this volume)

Vergnes M, Marescaux C, Micheletti G, Depaulis A, Rumbach L, Narter JM (1984) Enhancement of spike and wave discharges by GABA mimetic drugs in rats with spontaneous petit mal-like epilepsy. Neurosci Lett 44: 91–94

Vergnes M, Marescaux C, Depaulis A, Micheletti G, Narter JM (1987) Spontaneous spike and wave discharges in thalamus and cortex in a rat model of genetic petit mal-like seizures. Exp Neurol 96: 127–136

Wojcik WJ, Neff NH (1983) γ-amino butyric acid B receptors are negatively coupled to adenylate cyclase in brain and in the cerebellum of these receptors may be associated with granule cells. Mol Pharmacol 25: 24–28

Wood JH, Hare TA, Glaeser BS, Bellenger JC, Post RM (1979) Low cerebrospinal fluid γ-amino butyric acid content in seizure patients. Neurology 29: 1203–1208

Authors' address: Prof. N. G. Bowery, Department of Pharmacology, School of Pharmacy, 29-39 Brunswick Square, London WC1N 1AX, United Kingdom

Subject Index

H. Bönisch, K.-H. Graefe, S. Z. Langer, and E. Schömig (eds.)

Supplementum 34

Recent Advances in Neuropharmacology

The book presents the proceedings of a symposium held in honour of Professor Ullrich Trendelenburg. It is concerned with current research and provides new information on the pharmacology and physiology of such neurotransmitters as noradrenaline, dopamine, 5-hydroxytryptamine and ATP. The articles have been written by leading experts in the field of transmitter pharmacology. The most important feature of the book is that it brings together prominent scientists from a wide range of research areas. The new research activities and results then report on advance our understanding about the autonomic nervous system. Since the book provides the latest information on various aspects of neuropharmacology, it will certainly be of interest to a great many readers and enable them to bring their knowledge up-to-date.

Springer-Verlag
Wien New York

1991. 44 figures. VIII, 221 pages.
Soft cover DM 120,-, öS 840,-
Reduced price for subscribers to
"Journal of Neural Transmission":
Soft cover DM 108,-, öS 756,-
ISBN 3-211-82300-X

Prices are subject to change without notice

Supplementum 33

L. Deecke and P. Dal-Bianco (eds.)

Age-associated Neurological Diseases

This volume covers selected up-dated contributions from the Conference on "Age-associated Neurological Diseases" held in Vienna 1989. Fields covered include Age-associated Memory Impairment (AAMI) Alzheimer's and Parkinson's diseases and dementias and movement disorders of other etiologies.

Concerning dementia, some papers deal with diagnosis employing neuro-imaging methods–PET, SPECT, MRI and XCT, others by means of electro-physiological methods. An important aspect is the early preclinical diagnosis of dementia using neuro-psychological tests, to enhance the chance of effective treatment.

Finally, drugs now under clinical investigation are discussed and preliminary results for several compounds are presented. This volume with its up-to-date contributions will be of special interest to all physicians treating elderly persons with Age-associated degenerative diseases.

Springer-Verlag
Wien New York

1991. 30 figures. VIII, 165 pages.
Soft cover DM 98,–, öS 690,–
Reduced price for subscribers to
"Journal of Neural Transmission":
Soft cover DM 89,–, öS 621,–
ISBN 3-211-82261-5

Prices are subject to change without notice.